D1333388

014390036 3

Jen Chillingsworth

GROW GREEN

Tips and advice
for gardening
with intention

Illustrations by Amelia Flower

Hardie Grant

QUADRILLE

Publishing Director Sarah Lavelle
Senior Commissioning Editor Harriet Butt
Copyeditor Gillian Haslam
Series Designer Gemma Hayden
Illustrator Amelia Flower
Head of Production Stephen Lang
Senior Production Controller Katie Jarvis

Published in 2021 by Quadrille, an imprint
of Hardie Grant Publishing

Quadrille
52–54 Southwark Street
London SE1 1UN
quadrille.com

Cataloguing in Publication Data: a catalogue
record for this book is available from the
British Library.

Text © Jen Chillingsworth 2021
Illustrations © Amelia Flower 2021
Design and layout © Quadrille 2021

ISBN 978 1 78713 5727

Printed in China

Contents

Introduction

With rising temperatures across the globe and the reality of climate change now hitting home, it is more important than ever that we consider our actions in every area of life. This includes gardening – an activity we tend to think of as inherently 'green'. However, the energy consumption of power tools and lawn mowers, the manufacturing and transportation of horticultural products and the overuse of artificial fertilizers can all have a negative environmental impact on the planet.

There are many simple ways we can make a difference. Harvesting rainwater, choosing peat-free potting compost, growing our own food and providing a home for wildlife are all good ways to become a more eco-conscious gardener. Digging a little deeper, there are lots of other ways we can utilize our gardens or skills to be greener, such as growing cut flowers instead of buying imported, drying herbs to use in cooking, upcycling junk into planters or joining a community garden project.

Urban gardens play a vital role in helping to combat the problems of climate change. Planting trees and bushes can help to prevent rising levels of air pollution and keep temperatures cooler. If there are periods of intense rain, plants and trees can slow the runoff of water and reduce the pressure on urban drainage systems. They can also offer homes to birds and other wildlife and provide bees and other pollinating insects with a much-needed food supply.

A green gardener is essentially a thrifty gardener, buying less and choosing second-hand. You use up what you have and find ways to reduce your waste. From repurposing household plastics,

making your own compost and turning food waste into liquid fertilizer to feed growing plants that have multiple uses for attracting pollinators and deterring pests.

At home, I have a tiny courtyard-style garden where I grow edibles, flowers, herbs and climbers. I love watching how this garden changes through the seasons, from the calm, soothing greens and gentle colours of spring to the warming reds and rust shades of autumn (fall). In early spring, my neighbour's cherry tree flowers and turns the entire garden a bright shade of pink, with the merest hint of wind creating a shower of blossom.

In summertime, the rickety fence is engulfed by clematis, honeysuckle and jasmine. Old terracotta pots are filled with wildflowers and perennials, junk watering cans overflow with ginger, mint and sage, sheltering our resident hedgehog's house from potential predators. Chamomile 'Treneague' springs up between the cracks in the paving, releasing its intoxicating scent when gently stepped on. Peas and beans in a dizzying array of colours wind their tendrils around bamboo canes, whilst bees and butterflies forage for nectar on salvias and echinacea. I cut bunches of poppies, wild carrot, cosmos and sweet peas to bring into the house, filling earthenware pots and flea-market vases to decorate my mantelpiece. Strawberries, raspberries and blackberries are turned into jams and preserves. Tomatoes are canned and made into pasta sauces.

When autumn beckons and the leaves from neighbouring trees turn golden brown and fall to the ground, I rake them up and add them to the compost heap. The hedgehog curls up in her cosy home to hibernate for winter. Robins, blackbirds and squirrels visit the feeders, feasting and gathering supplies to see out winter.

This tiny garden is my happy place. I can sit at the rusty table with a cup of coffee for hours, listening to the birds chirping in the trees and simply watching nature do its thing. I don't notice the sound of the traffic on the roads nor do I feel the need to check my phone for notifications. It's an escape, an oasis in the middle of a busy city.

You don't need a big garden to grow things, in fact you don't even need to have a garden! A couple of pots by your front door or on a balcony, a window box or even a small seed tray on a kitchen windowsill can all produce wonderful results. There are lots of ways to garden inside the home – vertical plant walls, bokashi composting, microgreens and mushroom kits to name just a few.

This book is an introduction to eco gardening, from making your own fertilizer with leftover food, reducing energy consumption, conserving water and growing organically, to dealing with pests and diseases. It will hopefully inspire you to think more about

biodiversity and how to introduce native species and encourage more wildlife into the garden. I also hope it will make you look closer at what you buy in the garden centre or grocery store and with a little work, you could be growing flowers, fruits, herbs and vegetables at home in a more sustainable way.

Grow Green
Essentials

Grow green essentials

This chapter looks at the small changes that can make our gardens more sustainable. Many of these involve modifying our habits so they become part of our daily lives.

A good place to start is with the five guiding principles: refuse, reduce, reuse, recycle and rot. By applying these principles to the garden, you become more mindful of resources and reduce the amount of waste you produce. Before buying any equipment or tools, think whether you can repurpose or borrow an item. Do you have a local plant nursery or lumber yard who can supply you rather than travelling to a large DIY store or garden centre?

Single-use plastics cause huge damage to the environment, so reusing or repurposing plastics can offer a practical solution to the problem. In the garden, household items such as plastic milk bottles and cleaning sprays can be turned into vessels for watering whilst egg boxes and plastic fruit punnets are useful for seed sowing.

One of the most impactful ways to help the planet is to use less water. In warmer weather our water consumption soars, forcing suppliers to use groundwater and pump water from rivers and streams. This causes harm to the wider environment and threatens animal habitats. By making a few simple changes, such as growing drought-resistant plants or using moisture-retaining containers, you can significantly reduce the amount of water you use.

I have also shared other ways to help make your garden more sustainable. From planting trees or hedges that reduce air pollution to choosing hard landscaping materials, each plays an important role in creating a green garden.

Essential equipment

If you are new to organic gardening or want to make some changes to the way you already garden, one of the best ways to start is to look at the equipment you have or need to buy. When choosing tools, buy the best you can afford as these will be of a higher quality and should last longer. Test garden tools for comfort and size before purchasing, too. Condition any wooden handled tools by treating them with a drop of linseed oil on a cloth, and if the handle is damaged look online for a replacement. Car boot sales, garage sales, junkyards and vintage and salvage fairs are great places to look for used tools and watering cans if you don't want to buy new.

Compost and potting mixes should also be a consideration. Many multi-purpose composts contain peat, a non-renewable resource which is extracted from peatbogs (peatlands). Peatbogs are of great importance to our planet as they are capable of absorbing and storing large amounts of carbon

dioxide and provide vital habitats for rare plant species and wildlife. (On average peatbogs store 10 times more carbon per hectare than any other ecosystem including forests*.) Look for certified peat-free composts and environmentally friendly alternatives made from coir (coconut fibre), green matter or wool.

It's also a good idea to switch from commercial garden products to natural ingredients that you can use to make your own non-toxic pesticides (see pages 160–171). The following pages list my garden essentials, from planet-friendly gardening gloves and watering cans to bicarbonate of soda (baking soda) and ground cinnamon – all of these will help you to garden more sustainably.

* Garden Organic.
www.gardenorganic.org.uk

TOOLS

For gardening outdoors you will need a hand trowel, hand fork, hoe, rake, secateurs (pruning shears), scissors and a spade. Indoor gardeners will need a hand trowel and fork as well as secateurs.

GLOVES

Look for gloves made from natural rubber and organic cotton rather than ones made using fossil fuels and pesticides. You can find these online at plastic-free or fair-trade stores.

WATERING CAN

Choose a metal watering can as it will generally have a much longer lifespan than its plastic alternative. A watering can with a detachable rose is best, as you may need to

continued »

adjust the flow, depending on the plants you are watering. However, watering cans made of metal can be heavy to carry and aren't practical for everyone, so a good alternative is to buy a recycled plastic watering can or make your own from household plastics (see page 19).

GARDENER'S SOAP

Choose a good-quality bar of soap to wash your hands after gardening. Ideally handmade and vegan, look for bars that are made with pure essential oils, responsibly sourced oils or butters and contain poppy seeds or coffee grounds to help with exfoliation and removing ingrained dirt.

DISTILLED WHITE VINEGAR

This has antiseptic and antifungal properties and is used for testing the pH level in the soil. You can buy a glass bottle of distilled white vinegar (it may also be known as distilled malt vinegar) from the supermarket or grocery store or buy in bulk from online stores.

BICARBONATE OF SODA (BAKING SODA)

A natural fungicide used in the garden for cleaning, pH soil testing and to make pesticide. Buy loose from refill stores or in bulk from online stores.

LIQUID CASTILE SOAP

Completely biodegradable, vegan and safe for use around pets. Made from natural vegetable oils, liquid castile soap is used to make insecticide and treat powdery mildew. I refill my own bottle at a local zero waste store so search online for refill stores near you. You can also find liquid castile soap at your local health food store and major online stores. Be sure to choose castile soap that is free from palm oil or uses certified sustainable palm oil as an ingredient.

NEEM OIL

A natural, organic and biodegradable pest repellent and anti-fungicide. Safe to use around pets and wildlife, neem oil is extracted from the seed of the neem tree and it has been used for centuries as a natural pesticide. It will kill pests on plants, but if used correctly will not harm pollinating insects. Buy organic neem oil from pharmacies and online retailers.

PEPPERMINT ESSENTIAL OIL

Used for repelling garden pests such as aphids and whitefly and as an anti-fungicide. Peppermint essential oil can also help to deter slugs and snails from the veg patch. Look for organic essential oils and avoid using cheaper oils as they are often diluted and have not been sustainably produced.

GROUND CINNAMON

You may already have a jar of this in your kitchen cabinet for baking, but ground cinnamon also works as a natural fungicide in the garden. Sprinkle it lightly over the top of the soil in seed trays as it helps to prevent 'dampening off' (a fungal disease that affects seed growth). It also works well as a rooting powder for cuttings and deters pests. I like to buy ground cinnamon at my local zero waste store as I can fill my own container. Many online stores sell cinnamon in bulk which reduces the need for packaging and is more cost effective if you propagate a lot of seeds or cuttings.

The 5 Rs for growing green

Whenever you make want to make a sustainable change to your home or garden, it's important to consider the 5 Rs: refuse, reduce, reuse, recycle and rot. By applying these guiding principles, you can make better choices for your household and the planet.

Here are a few ways you can use the 5 Rs to help you grow green:

REFUSE
To buy products that will ultimately end up in landfill. These are usually made from plastics that can't be recycled and include netting, plant labels and plastic plant pots. Try not to buy plants that have travelled vast distances to be sold at large DIY chains or supermarkets. Instead, look for plants that have been grown in local nurseries as less energy and transport are required to produce them and you'll often find more unusual varieties on sale, too.

REDUCE
Learn to stop making impulse purchases and only buy what you need.

Rather than buying equipment that you don't use regularly, hire it. Lawn mowers, hedge trimmers and chainsaws can all be rented from tool hire companies for a small daily charge. Alternatively, check online to see if there is a tool library in your area. These are social enterprises that promote sharing as a way of reducing your environmental impact. For a small annual membership fee, you can then access a huge selection of gardening and DIY tools.

REUSE
Maintain your gardening tools by cleaning them after each use and rubbing them down at the end of the season with wire wool (this helps to prevent rust build-up). Look online for tutorials that teach

you how to sharpen blades on spades, hoes and pruning shears. If a small appliance like a hedge or grass trimmer breaks down, rather than replacing it, take it to a repair café run by volunteers who will help you to fix it for a small contribution. Look online for a repair café near you.

Keep hold of your empty potting compost bags to line raised beds (punch a few drainage holes first) or to use as a grow bag for potatoes, peas or broad (fava) beans. If you buy potting compost from a local garden centre or nursery, ask if you can return the bags for them to reuse.

Many recycling centres offer their own potting compost produced from collected garden waste. Search online to find out if your local authority offers this service. If you have a bumper harvest, donate your excess produce to a local food bank or a social enterprise project to help benefit your community. There are lots of everyday household items such as plastic bottles,

egg boxes and fruit punnets that can be reused in the garden too. See pages 19, 20–23 and 63 for ways to repurpose these.

RECYCLE

Whatever you can't refuse, reduce or reuse. Check online to see what items your local authority will accept for recycling.

ROT

Compost whenever possible, but if you don't have room for a compost bin outdoors there are several indoor alternatives and other ways you can make compost from food and garden waste. See pages 46–55 for further details on composting.

Water-saving ideas

One of the best ways we can grow green is to reduce our water consumption. There are several ways you can do this before you even pick up a watering can:

ORGANIC MATTER

Adding homemade compost or leaf mould to the soil before planting will improve its ability to retain moisture (see pages 34–37 for further details).

MULCHING

Adding a layer of mulch to the soil helps prevent water evaporating and reduces the frequency for watering.

WEEDING

Weeds steal nutrients and water from plants. Regular weeding helps prevent this from happening.

CONTAINERS

Avoid using terracotta pots for planting as they are porous and will lose water. Choose glazed pots instead. Pots or raised beds made from metal will heat up quickly and draw more moisture from the soil, so will require more watering. If plants have become root-bound they will require more watering, so repotting them into a large container will also reduce water consumption.

DROUGHT-RESISTANT PLANTING

If you live in an area that has low rainfall or you simply want to cut down on watering, choose plants that are adaptable to dry conditions. These are usually plants with leaves that are furry, waxy, grey or silver. Agave, lavender, rosemary, rudbeckia (Black-Eyed Susan) and thyme will all thrive without water for long periods of time.

WHEN TO WATER

Water plants only when they really need it rather than giving them a light watering every day.

A good soaking every few days helps plants develop healthier root systems. Prioritize watering hanging baskets, containers and any new plants over more established specimens. And if possible, water in the early morning before the temperature rises or in the early evening.

WATER THE ROOTS

Plants soak up water from the soil, so aim your watering can directly at the roots and not the leaves.

Equipment for watering

» *Watering can:* Invest in a good-quality watering can with a removable rose (see page 13). You can also make your own watering can from plastic milk cartons or laundry bottles (the large sizes, which have a handle). Simply wash the bottle in warm soapy water, rinse, then leave to air dry. Remove the screw-on lid from the bottle and, using a small drill bit, drill multiple holes in the cap. Fill the bottle with water, reattach the lid and water as normal.

» *Hosepipes:* If you use a hose to water in the garden, choose one with a nozzle that will close off the water supply. This prevents water from flowing freely whilst you go to turn off the tap (faucet). Avoid using sprinkler attachments as these consume vast quantities of water and are not always effective.

» *Drip irrigation:* Installing a drip irrigation system is a great way to conserve water and works for containers, flower and vegetable beds, raised beds, window boxes and indoor plants. This system uses a combination of hoses and drippers, delivering water slowly to the root of plants. It's inexpensive and easy to install, saves both money and time and can help to prevent diseases such as powdery mildew and blight, which can be caused by water sitting on the surface of the soil.

» *Water butt:* Choose the largest size of water butt you can fit in your garden and, if you have the space, invest in a few and attach them to the downpipes from your roof guttering. It's worth noting that water butts can get mucky

continued »

and the water often turns stagnant. To counter this, use a biodegradable and non-toxic water butt cleaning product. Always secure the lid on the water butt to prevent flies and mosquitos from entering and laying eggs (and also to ensure that inquisitive animals or small children cannot topple in).

» *Ollas:* These are large pots or bottles made from terracotta that are buried in the soil, leaving the top or neck sitting above the ground. They are filled with water and any plants surrounding the olla will draw moisture from the pot towards their roots. You can buy ollas at garden centres and nurseries, or it's easy to make your own with unglazed terracotta plant pots. Start by filling the drainage hole in the plant pot with either a little mounting putty or use an old wine cork. Bury the pot in the soil near any plants that need watering, leaving the top of the pot sitting above the ground. Fill the pot with water, then cover with a terracotta or plastic saucer to prevent evaporation. Check the pots regularly and refill when necessary.

Recycle water

» *Rainwater:* On days with heavy rainfall, place your watering can (and any other gardening tubs you may have) outside to catch droplets. Attaching a funnel can help direct the rainfall into a watering can. This water is ideal for watering indoor gardens and houseplants, which don't often get the benefit of natural weather conditions.

» *Cooking water:* Set aside water that has been used to cook vegetables, pasta or rice. Let it cool to room temperature before adding it to your watering can and use it to water indoor and outdoor plants.

» *Grey water (graywater):* Used water that has come from your home can be collected to water both indoor and outdoor plants. However, there are a few general rules to apply before using grey water. Firstly, only water that has come from dish washing or sinks, baths and showers should be used (not any toilet waste water). Don't use any water that has had commercial cleaning products added as these can leach damaging chemicals and non-biodegradable materials into the soil. Instead, choose ecological and plant-based cleaning products that will do no harm. Finally, allow the water to cool to room temperature before using and apply to the roots of the plants and not the leaves when watering. Alternate with stored rainwater for the best results.

» *Dehumidifiers:* Water produced in a dehumidifier is considered safe to use in the garden. Simply pour the contents of the water tank into a watering can and use on indoor and outdoor plants. However, water in the tank can contain toxins so it's best to avoid watering any plants that are edible. Dehumidifier water should be treated in the same way as grey water and alternated with stored rainwater for the best results. Be mindful of keeping the water tank clean to avoid any contamination from mould.

Going green in the garden

We are all aware that plastic production is harming our planet. Sadly, there is a vast array of garden products packaged or made from plastic, but we can significantly reduce our consumption.

Some plastics can be reused safely in the garden as they won't break down quickly or leach toxins into the soil. All plastics have a number embossed on them, usually inside a small triangle, and these numbers relate to the type of plastic that the product is made from. Although many people recommend using soft drinks bottles in the garden, plastics numbered 1 (polyethylene terephthalate) are generally regarded as single-use plastics that degrade quickly in light and heat and can contaminate the soil. It's therefore safer to stick to plastics numbered 2 (high-density polyethylene) or 4 (low-density polyethylene) as these resist UV rays and can tolerate high temperatures. These include

milk bottle jugs (pitchers), food containers and detergent bottles.

PLANT POTS AND TRAYS

These are usually made from recycled plastics, but in most cases can't be recycled at the end of their lifespan. If you can't reuse a plant pot, ask the garden centre or nursery where you bought the plant if they can reuse the pot or tray. For seedlings, choose biodegradable pots made from coir or make your own from old newspaper (see page 64). Yogurt pots and egg cartons also make fantastic pots to start off seedlings.

PLASTIC MILK BOTTLES

See page 19 for instructions on how to make a watering can from a plastic milk bottle. These bottles can also be repurposed for seed growing. Use a marker pen to draw a line around the bottle approximately one third of the way up from the bottom. Using a strong pair of scissors or a crafting knife carefully cut around the line so you have two pieces of plastic. Wash with warm, soapy water, rinse and leave to air dry. You can use the bottom, flatter part as a seed tray by punching a few drainage holes in the bottom.

CLOCHE

Following on from above, the top part of a plastic milk bottle can provide protection for young plants. Simply place over the plant and push the edges into the soil. If the weather is warm, remove the plastic bottle cap to allow air to circulate (keep hold of the bottle cap as you will need to reattach it if the temperature drops). Windy weather can blow the bottle away and you may need to attach some wire pegs to the bottom to give added support.

PLANT LABELS

Wash and set aside wooden ice lolly (popsicle) sticks or bamboo toothbrush handles (bristles removed) to use in place of plastic plant markers. Keep hold of wine bottle corks (not plastic ones),

write the plant variety on the cork with a permanent marker pen, then attach the cork to a bamboo skewer by inserting the point of the skewer into the thin end of the cork.

NETTING

Switch to unwaxed natural fibre twines made from jute, sisal or hemp and make your own garden netting. Search online for tutorials to create pea and bean supports, fruit cages and trellising. At the end of the growing season, add the used twine to the compost bin.

SPRAY BOTTLE

For misting seedlings, rather than buying a new spray bottle and mister, it's a more sustainable option to repurpose a bottle you may already have in your home. Bottles that once contained non-toxic cleaning products are fine but avoid using those that once contained commercial cleaning products. No matter how well you wash these, there is always the risk that some residual residue may remain, and this can cause harm to plants and pollinating insects.

continued »

Other ways to go green in the garden

IMPROVE AIR QUALITY IN YOUR NEIGHBOURHOOD

» *Plant a native tree or hedge:* Single trees in urban areas have been found to be far more beneficial at collecting harmful particles than those in a woodland area. Planting low hedges in urban areas can help to reduce the impact of pollution as they are closer to the level of vehicle exhaust fumes and can absorb the particles before they have the chance to disperse into the air. Search online to find native trees and hedges for your area.

» *Green roof:* Installing a green roof in an urban environment helps to filter out airborne particles and pollutants and reduce noise pollution. It consists of a waterproof layer, a root barrier, a drainage system and a growing medium for the plants. In winter, a green roof will insulate against heat loss, and during the summer it will retain cool air. They are usually planted with low-maintenance sedums, succulents, herbs, wildflowers or grasses, but you can also grow salad crops and strawberries. A green roof is a great way to attract pollinating insects to your garden.

If you cannot install a green roof on your house, then sheds, garages or garden buildings are a great alternative. You can purchase prefabricated green roof buildings or search for online tutorials on how to convert one. Log stores, dustbin (trash can) cupboards, bike storage and kids' playhouses are other good options.

» *Go vertical:* Perfect for balcony gardens, fence panels or a wall, growing vertically makes use of otherwise dead space. Lots of herbs, flowers and vegetables can be grown successfully this way. Create your own DIY versions from upcycled pallets, repurposed bookcases or wooden ladders or by attaching lengths of guttering. There are many DIY video tutorials online, or alternatively, you can purchase modular kits to get you started.

CHOOSE SUSTAINABLE LANDSCAPING MATERIALS

» *Reclaim:* Consider using reclaimed materials such as house bricks for paths and timber sleepers (railroad ties) or pallets for raised beds.

» *Go natural:* Use natural materials such as bark or wood chippings for pathways, and ask a local tree surgeon if you can buy directly from them. Seashells are another good option – as a by-product of the shellfish industry, they can be used as a mulch and to make pathways (don't collect from the beach). Rather than using cement to fill in gaps between paving stones, add sand or create a living patio: plant low-maintenance ground-cover plants between paving, such as creeping thyme or chamomile 'Treneague' which will attract pollinating insects.

» *Selecting wood:* Check any wood you use is FSC (Forest Stewardship Council) or PEFC (Programme for the Endorsement of Forest Certification) registered, as this ensures that the wood has been sourced from a forest managed in line with strict environmental, social and economic requirements.

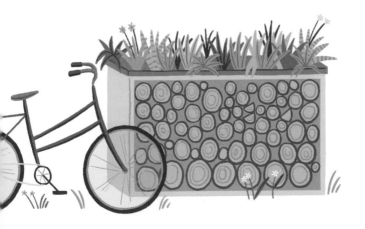

Soil & Fertilizers

Soil & fertilizers

When I first started gardening, I really didn't know what I was doing. I simply bought plants that I thought were pretty, but sadly many of them struggled as I didn't provide them with what they really need to flourish – healthy soil packed full of nutrients.

A healthy soil has good structure, retains moisture and delivers nutrients to plants, so spend a little time researching the soil in your garden. You may be lucky and find it to be healthy and productive, but chances are it's lacking something. There are two main things to consider: soil type and the pH level. Both can be determined by collecting soil samples from your garden and carrying out some simple tests. The results from these tests will help you plan what you can grow or what nutrients are lacking and, if necessary, make amendments to improve the soil.

Soil improvers are made from organic matter and are either dug into the soil or spread over the surface as a mulch. On pages 34–37 you'll find lots of different ways you can add organic matter, from seaweed gathered on the beach or grass clippings to leaf mould and mushroom compost. They are all environmentally friendly, inexpensive and effective.

Plants should get everything they need from the soil if it is healthy. However, challenging weather conditions and soil erosion can lower levels of vital nutrients and plants may need to be given a boost. It's easy to make your own organic fertilizers from plants or weeds already growing in the garden or even from food waste. Nettles, comfrey, eggshells and coffee grounds are all rich in essential vitamins and can easily be made into concentrated liquid feeds (see pages 38–43).

How to determine soil type

Many plants won't grow properly if the soil isn't right for them, so it's vital to find out what type of soil you have in your garden before you do anything else.

Follow the steps below to determine your soil type. It's a good idea to take a few samples from different areas of the garden as soil type can vary within your plot. For a more detailed analysis, send a sample to a specialist soil-testing laboratory. These are widely available and can be found online.

HOW TO TEST FOR YOUR SOIL TYPE

Take a handful of soil from your garden and add a little water. Place it in one hand and, using the other hand, squeeze the soil together to form a ball shape. When you open your hands, the soil ball will do one of the following:

» hold its shape well, indicating clay soil

» roll into a ball but won't hold its shape for very long, indicating loam soil

» feel spongy if squeezed, indicating peat soil

» immediately fall apart, indicating chalky or sandy soil

» have a slippery texture and not clump together, indicating silt soil.

TYPES OF SOIL

» *Chalk:* Alkaline, stony and free draining. Not suitable for growing acid-loving plants as the soil is highly alkaline. If you turn over the soil lightly with a trowel, you will see tiny lumps of chalk on the surface.

» *Clay:* Warms up slowly in spring, turns hard and cracked when dry and drains slowly. Avoid planting in spring and focus on improving the soil over the autumn (fall) and winter months.

» *Loam:* Warms up quickly in spring and is the ideal soil type as it is packed full of nutrients, is free draining and does not get waterlogged.

» *Peat:* Rarely found in gardens but good for growing acidic-loving plants such as azaleas, blueberries and rhododendrons.

» *Sandy:* Warms up quickly in spring and is free draining. Add plenty of organic matter to boost nutrients and increase the water-retaining capacity.

» *Silt:* Higher in nutrients than sandy soil, is free draining and retains moisture well. However, the fine particles can be easily compacted and are prone to washing away in heavy rain.

In most cases, soil can be improved by adding organic matter (see pages 34–37). However, with certain types of soil you may need to reconsider what you plan to grow. Acid-loving plants such as blueberries and azaleas will not do well in chalk soils, whereas gardens with peat soil will have ideal conditions. If you find yourself with chalk soil, plant acid-loving specimens in containers with ericaceous potting compost.

HOW TO IMPROVE YOUR SOIL TYPE

» *Chalk:* Add plenty of organic matter. Mulch well (see page 34) and add green manures such as clover or vetch.

» *Clay:* Add plenty of organic matter (see page 34) and spread on top of the soil. Using a spade, mix the organic matter into the top 20cm (8in) of soil. Add new plants in spring and avoid planting anything new in the autumn as it may not survive a waterlogged clay soil over winter.

» *Sandy:* Apply organic matter regularly and mulch well to help with moisture retention. Adding a layer of slate chips, gravel or pebbles also helps sandy soils retain water.

» *Silt:* Add a 5–10cm (2–4in) layer of organic matter over the surface of the soil in early spring or autumn. Fork in or leave it for the worms to incorporate it into the soil.

Understanding the pH level of soil

Plants should get most of their nutrients from the soil, but if it is too acidic or too alkaline, those nutrients won't be available to them. Before you begin planting in a new bed, it's a good idea to test the pH level and see if you need to make any amendments to the soil. If you have established vegetable or flower beds and your plants are not growing well, it's worth investigating the soil's pH level as this may be the underlying cause.

You can buy pH testing kits from garden centres or send a sample to a specialist laboratory, or you could do a general test using ingredients you may already have at home (see right). However, a testing kit will give you a more accurate result and make subsequent soil amendments easier. Soil samples sent to specialist laboratories go further, giving you a breakdown of the nutrients and organic matter in the soil. Search online for soil sample analysis services.

Most plants and vegetables prefer soil that has a pH level between 6 and 7.5, in the neutral range of the scale. Neutral pH levels mean essential nutrients like nitrogen, phosphorous and potassium are easily dissolved and delivered quickly to plants. Once you know the soil's pH level, you may need to make amendments to achieve a pH level between 6 and 7.5.

ACIDIC SOIL (LESS THAN PH7)

» *Lime:* Spread or incorporate into the soil. Allow 2–3 months for the lime to neutralize the acidity of the soil. Find online or in garden centres.

» *Biochar:* Dig into the soil (see page 34 for further info).

» *Wood ashes:* Spread ash from bonfires or wood-burning stoves directly onto vegetable beds in late winter, then fork or dig in. Avoid using wood ash where you grow potatoes as very alkaline soils can encourage potato scab.

ALKALINE SOIL (MORE THAN PH7)

» *Compost:* Add mature compost to topsoil and spread over planting areas. (See pages 48–53 for how to make compost.)

» *Sulphur chips:* Mix well into the soil or scatter over the surface. Find online or at garden centres.

You may have to reapply lime or sulphur chips every year to keep your soil pH neutral. Apply in autumn (fall) or after the growing season to allow the adjustment to work effectively.

Note that certain plants, such as blueberries, rhododendrons, camellias and heathers, thrive in more acidic soil. Rather than making amendments to the soil's pH level, try planting varieties that love those acidic conditions.

Testing soil's pH

YOU WILL NEED
» Trowel
» Two containers
» Distilled white vinegar
» Water
» An old spoon
» Bicarbonate of soda (baking soda)

1. Dig two samples of soil and place in separate containers.

2. Pour 250ml (½ cup) distilled white vinegar over one soil sample. If it bubbles or fizzes, it is likely to be alkaline with a pH of 7–8.

3. For the second sample, pour over 250ml (½ cup) water and stir. Add 90g (½ cup) bicarbonate of soda. If the soil bubbles or fizzes, it means it is likely acidic with a pH level of 5–6.

4. If you do not get either of these reactions, the soil is neutral.

Organic matter & mulching

All plants benefit from growing in healthy and good-quality soil, but many of us struggle to provide them with this. We may have clay soil, which can drown root systems, or sandy soil, which can drain water away quickly and deprive plants of moisture and air (see pages 30–31 for identifying soil types). However, we can easily improve the structure of our soil by adding organic matter and mulching.

Organic matter is dug into the soil or used as a mulch. Adding it to the soil will increase the soil's fertility and feed essential microbial life. It is not a fertilizer, but some options release nutrients into the soil.

Mulching is used specifically for weed suppression and to aid moisture retention. Weed planting areas thoroughly and water well before adding any organic mulches.

Here is my list of options for adding organic matter or mulching. Some are more accessible than others, but the majority are easy to make at home and are either relatively inexpensive or free. Although it is a good environmental option, I have chosen not to include cocoa bean/shell mulch on this list as it is toxic to both cats and dogs.

BIOCHAR
Improves soil structure, retains nitrogen and phosphorous in the soil, benefits plant growth and root development. Biochar is created by the slow burning of plant material with little or no oxygen. Dig biochar into the soil and it will increase crop yield, break down pesticides and even decrease greenhouse gas emissions. Find biochar online from specialist suppliers and organic gardening companies.

CARDBOARD AND PAPER

Used as a mulch for weed suppression. Lay sheets of cardboard or paper over areas to block the light and kill weeds. The cardboard or paper will decompose over several months and worms will also break it down, leaving the area ready to plant. Plain paper (white or coloured), newspaper, plain and non-shiny cardboard can all be used for this purpose.

COMFREY

Plant a separate bed with comfrey (see page 39). Comfrey adds nitrogen, phosphorous and potassium into the soil as the leaves decay. Dig leaves into the soil before planting or use as a mulch by simply placing a layer of comfrey leaves 5cm (2in) thick on the surface of the soil around all plants.

GARDEN COMPOST

Add compost to your flower and vegetable beds to improve soil structure and attract earthworms. Dig in or add as a mulch. (See pages 46–53 for how and what to compost.)

GRASS CLIPPINGS

These add nitrogen, phosphorous and potassium into the soil. Be sure to only add grass clippings that have not been treated with pesticides or herbicides. Apply as a mulch after weeding, spreading thinly over the surface of the soil. After a couple of days, apply another thin layer and after a week, add a final thin layer.

GREEN MANURE

This is a temporary crop that is allowed to grow, then broken down and added to the soil. Green manures improve soil structure, add valuable nitrogen to increase the fertility of the soil and suppress weeds. Crimson clover, mustard, rye and vetch are all excellent green manure crops. Sow seeds from spring through to autumn (fall). Crops will need to be 'dug in' at least four weeks before you want to plant in that area.

LEAF MOULD

Collect leaves from deciduous trees in the autumn. Gather from your garden, the local park or ask a neighbour for their leaves.

continued »

Don't collect leaves from woodland areas – leave these for wildlife. Place in a biodegradable jute leaf composting bag (find them online) and moisten the leaves with water. It will take about a year for leaves to decompose. Once they do, dig the leaf mould into vegetable beds or use as a mulch. Leaf mould doesn't add a lot of nutrients to the soil, but it is a great conditioner as it increases water retention and provides a valuable habitat for earthworms. Gather pine needles separately as they produce an acidic leaf mould, ideal to mulch acid-loving plants like blueberries and camellias.

MUSHROOM COMPOST

A by-product of mushroom farming, this can be dug into the soil and used for mulching, but avoid using around acid-loving plants such as blueberries, rhododendrons and azaleas. It is an excellent source of phosphorous, nitrogen and potassium. You can find mushroom compost online.

SEAWEED

Freshly gathered seaweed works wonders in the garden. As it breaks down in the soil, it releases nutrients and hormones to encourage plant growth.

Seaweed doesn't clump or blow away nor does it contain any seeds that could grow into weeds. Slugs, bugs and birds don't like the smell, so it works well as a pest deterrent, too. Collect seaweed that has been washed onto the beach and not from rocks. Check with your local authority or whoever owns the beach before foraging for seaweed as some do not allow it.

SHREDDED PRUNINGS

Any pruning from shrubs or hedges can be used as a mulch. However, don't apply them to vegetable or annual flower beds as they can deprive the soil of vital nutrients. Use shredded prunings around mature shrubs and trees. Gather into a pile and leave for three to four months before spreading on the planting area.

COVER CROPS

If you have a bare vegetable, flower or raised bed, plant cover crops for overwintering. These suppress weeds, improve soil fertility, encourage biodiversity and prevent soil erosion. In late summer or early autumn (fall),

clear the beds of weeds and sow your cover crop seed. Clover and vetch are good cover crops as they pull nitrogen from the atmosphere and release it back into the soil, whilst mustard is beneficial for those with clay soils. Dig in cover crops in very early spring, ideally before crops have started to flower, or alternatively cut and leave on the surface for earthworms to work into the soil. Leave for at least one month before sowing or planting any new crops.

Organic fertilizers

Fertilizing is the last piece of the jigsaw puzzle for growing healthy plants. Fertilizers deliver extra nutrients, but they have no effect on soil structure or fertility. Adding a fertilizer helps plants to grow stronger, prevents diseases and leads to more productive harvests.

All fertilizers contain concentrated sources of plant nutrients. The main three are nitrogen (chemical symbol N, for leaf development and making your lawn green), phosphorous (chemical symbol P, aids root development) and potassium (chemical symbol K, for flowering, fruiting and all-round plant performance). Some plants also need nutrients in the form of trace elements such as calcium and magnesium, but these are usually only required in small amounts.

Organic gardening relies on fertilizers derived from plant or animal sources. They are slower acting than their manmade counterparts, yet they are more beneficial as they last longer and won't wash away when it rains. You can find commercial organic fertilizers at the garden centre, but it's just as easy to make a nutritious feed at home. I choose not to use any animal by-products in my garden, focusing instead on reducing the amount of waste I produce by utilizing food and plant materials.

Most homemade organic fertilizers can be made into liquid teas by adding water to them. These teas are best when made using stored rainwater, but if you haven't got access to a water butt, fill a bucket from the cold water tap (faucet) and leave for about 24 hours to allow any chlorine to evaporate.

BANANA PEEL

Rich in potassium and a good source of phosphorous and calcium, but they do not contain nitrogen. Place two banana skins into a large 1-litre (35-oz) jar and top with water. Cover the top of the jar with a cloth and leave for 48 hours. Remove the skins and add them to the compost heap. Apply the liquid immediately. You can pop banana skins in the freezer to make batches of this fertilizer when required. Use organic bananas where possible as they will not have been sprayed with harmful herbicides or pesticides.

COFFEE GROUNDS

A good source of nitrogen, phosphorous and potassium, plus other micronutrients. Dried coffee grounds are naturally acidic, and acid-loving plants will appreciate the occasional feed of dried fresh grounds. However, once the beans have been used, they have a neutral pH level of 6.5 and can be used to fertilize indoor and outdoor plants. Add 200g (2 cups) used coffee grounds to a large bucket filled with water. Leave overnight and in the morning transfer the liquid into a spray bottle to directly spray leaves.

COMFREY

Contains high levels of nitrogen, phosphorous and potassium, with many other trace elements. Comfrey is a wonderful plant to grow in an organic garden as it offers many benefits, but it does require some consideration before planting. It is vigorous and not

continued »

suitable for containers and, if allowed it can get out of control, so it's best planted in its own area. Buy sterile varieties of comfrey to prevent self-seeding. Use gloves for cutting the leaves as they can cause skin irritation. To make a liquid feed, follow the steps for nettle tea on page 43. Use undiluted for direct application. If you don't want to grow it or haven't the space, you can also find comfrey fertilizer online from natural gardening stores.

COMPOST

Contains nitrogen, phosphorous and potassium, as well as beneficial bacteria and microorganisms. To make a liquid fertilizer, fill an old pillowcase with compost (see pages 48–53 for how to compost), then tie the top with a piece of twine. Place in a bucket, fill with water and leave for two to three days. Remove the bag and empty the contents of the pillowcase back into the compost heap. Dilute the liquid that is left behind with fresh water in a ratio of 1:10 and apply immediately.

EGGSHELLS

A rich source of calcium and useful for feeding tomatoes and (bell) peppers. Collect any used eggshells, wash them and leave to air dry. Once you have a lot of shells, crush in a food processor until they resemble a fine powder and store in a glass container with a lid. It takes a long time for eggshells to break down completely in the soil and be absorbed by plants, so fertilize twice a year; ideally in the autumn (fall) when resting vegetable beds and then again in the spring. Sprinkle the eggshell powder around the base of plants. For pot-grown veg plants, add a little ground eggshell and organic matter to the bottom of the container before planting.

GRASS

Rich in nitrogen and phosphorous. Fill a bucket with grass clippings, weigh down with a brick, add water and leave for two days. Strain the grass into a second bucket and dilute the liquid in the ratio of 1:10. Use immediately or within two days.

VERMICOMPOST TEA

A nutritious compost tea made with worm castings (see pages 54–55 for how to create worm compost). Fill a piece of cheesecloth (muslin), a porous bag or an old pair of tights (pantyhose) with worm castings. Tie the top and place in a large bucket of water. Leave overnight and by morning the water will have turned a light shade of brown. Remove the bag and add the spent worm castings to the compost bin. Dilute the tea with water in a 1:3 ratio and apply within 48 hours.

SEAWEED

One of the best fertilizers to use in the garden as it contains potassium, magnesium and trace elements. Seaweed is completely sustainable and can be harvested without causing damage to the environment. Commercial liquid seaweed fertilizers are widely available, and you can also find dried seaweed and granules online to make your own seaweed teas.

TO MAKE FERTILIZER FROM FRESH SEAWEED

Gather seaweed that has been washed up on the beach, but always check you have permission from your local authority or beach owner before you do this. Take a large bag and use rubber gloves. When you pick the seaweed up, shake lightly to remove any hidden sea creatures or plastics (dispose of these safely, rather than leaving them on the beach). At home, place the seaweed in a bucket of fresh water and leave to soak for one hour to remove any salt.

For the next stage, you'll need to have two buckets, one large and one slightly smaller with holes in the bottom. Place the smaller bucket inside the larger bucket. Transfer the seaweed to the smaller bucket and fill with fresh water. Cover with a lid and leave for one month. Dilute the drained liquid from the smaller bucket with water in the ratio 1:5. Water the soil directly or dilute 1:10 for leaf spray application. Any seaweed remnants can be added to the compost heap.

Nettle tea fertilizer

Many people dislike nettles as they think of them as unsightly, troublesome weeds that give painful stings and so choose to remove them from their gardens. However, nettles offer us so much in the organic garden that it's useful to keep them around. They provide shelter and are a vital food source for insects, and if added to the compost heap they act as a natural activator to speed up decomposition. Nettles are particularly useful when broken down and turned into a nitrogen-rich fertilizer.

You can use nettles from your own garden or ask family, friends or neighbours for theirs. Be wary about collecting nettles from grass verges as they may have been chemically sprayed, and always wear protective long-sleeved clothing and a pair of gardening gloves when cutting nettles.

The best time to make this fertilizer is in spring, although nettles can be picked several times throughout the growing season. Each time you cut, add to the bucket and refill with water to make a new batch of fertilizer. Choose young nettles and avoid stems setting seed or flowering. Use this fertilizer on established leafy greens and avoid using it on seedlings as it's a little too strong for them.

How to make nettle tea fertilizer

YOU WILL NEED

- » Secateurs (pruning shears)
- » Gardening gloves
- » Nettle leaves
- » 2 large buckets (one with a lid or use a tray)
- » A brick or paving slab
- » Water (ideally rainwater)
- » Strainer (garden sieve or a piece of wire mesh)
- » Funnel
- » Container with lid (plastic milk carton or glass bottle)

1. Cut the nettles near the base of the plant and chop each stem into several pieces. Using gloved hands, crush the nettles to bruise them. Add to the bucket and keep filling until the bucket is nearly full.

2. Place a brick or paving slab on top of the nettles to weigh them down. Fill the bucket with stored rainwater, put on the lid or tray, and leave in a sheltered area of the garden or a greenhouse.

3. After three to four weeks, remove the lid or tray, (and stand back because it smells awful!). Pour the liquid through a large strainer into the second bucket. Attach the funnel to your chosen container and pour the strained liquid into the container.

TO USE

Always dilute nettle tea with water in a 1:10 ratio for watering plants and in a 1:20 ratio for spraying directly onto leaves. Apply once a fortnight throughout the growing season. Wear rubber gloves when applying the fertilizer as its smell is not pleasant and it will linger on hands and clothing.

Composting

Composting

Food waste sent to landfill can take a long time to rot and as it does so it releases methane, a gas more damaging to the environment than carbon dioxide. The transportation of waste requires vast quantities of fuel and energy, from trucks taking waste to landfill or from sending items overseas for other countries to recycle. We can drastically reduce what we send by setting up a composting system at home.

Composting outdoors is easy (see page 50). However, if you don't have a garden, there are several ways you can compost indoors. A DIY worm farm takes minutes to make (see page 54) or you can buy freestanding worm farms which are perfect for small spaces, indoors and out. Bokashi bins are also a brilliant way of composting food scraps indoors (see page 48). A food recycler machine heats up food scraps until dehydrated and sterile, then grinds them into tiny pieces that can be directly added to soil just four hours later.

If you don't want to compost indoors, look to your local community instead. Check with your local authority to see if they offer a kerbside collection for food waste. Ask your local school, allotment, community garden or urban farm if they can accept donations of food scraps for composting. Look online for websites and apps that locate local businesses who will take your food waste for free and turn it into compost. It's also useful to join a local zero waste group on social media as these are a great resource for lots of community projects.

There are many great organizations offering kerbside collection for homes, offices and restaurants. You will have to pay for this service, but it's usually inexpensive and they are often run by social enterprises.

Types of composting

Indoor composting

BOKASHI

Bokashi is a highly effective Japanese method of composting food scraps, ideal for those who live in small spaces. It is a fermentation process that uses a special bran inoculated with EM (effective microorganisms) to break down organic waste quickly. Bokashi bins and bran are available from specialist online suppliers. Bokashi works on all food scraps, including bones from meat and fish, cooked foods and dairy products.

The method is simple. Food waste is added to a bokashi bin, a tablespoon of the bran is sprinkled over the waste, the lid is replaced, and this is repeated until the bin is full. The bin is left for two weeks to complete the fermentation process. Every two to three days during this period, check for any excess liquid and drain if necessary.

This can be used to make a wonderful fertilizer for indoor and outdoor plants. Dilute 100 parts water to one part bokashi liquid. You can also add the bokashi liquid directly to kitchen and bathroom drains to prevent odours. After two weeks the food waste will have become nutrient rich. Bury it in the garden and leave it for 2–4 weeks before planting over it.

WORM FARMING (VERMICULTURE)

Worm farms convert organic waste into nutrient-rich fertilizer. The manure left behind from worms is known as worm castings or vermicompost and it's packed full of vital plant nutrients, minerals and microorganisms. Worm farms can be bought from garden centres and online, but it's easy to make one from an old dustbin (trash can), car tyres (tires) or wooden pallets.

You can find worms online at specialist worm farm suppliers or check out a local Freecycle group. Don't dig up earthworms from the garden as these are deep burrowing and they will not be happy living in a container. The most common type used for worm farming is the Tiger Worm (also known as the Red Wriggler). These worms can consume up to half of their body weight every day and reproduce quickly, making them the ideal choice for managing food waste.

Worm farms won't smell bad if looked after correctly. If they do start to smell, add more air holes, some shredded cardboard and reduce the amount of feeding until all food scraps have been processed.

Feed your worms once a week, burying the food waste in the bedding. The smaller your worm food is, the easier it is for them to eat. For fruits and vegetables, add them to a blender or food processor to make a 'worm smoothie' before adding to the worm bin. It will take several months before the worm castings are ready to be harvested. Search online for video tutorials on how to do this.

You can have a healthy worm farm outdoors, too. In the summer keep your worm farm out of direct sunlight, and in cold weather wrap your worm farm in old carpet or bubblewrap, or move it into a shed or garage.

WHAT TO FEED WORMS
» Fruits
» Vegetables
» Eggshells
» Coffee grounds and tea leaves
» Hair
» Toilet roll inners (shredded)
» Cardboard egg cartons (shredded)
» Corrugated cardboard (shredded)
» Indoor floral arrangements that have withered

WHAT NOT TO FEED WORMS
» Meat
» Dairy
» Onions
» Too many citrus fruits (they don't mind the flesh, but avoid the pith/skin)
» Fats or oils

continued »

Outdoor composting

There are two ways to make compost outdoors – cold composting and hot composting. Both have their advantages and disadvantages. Cold composting is simple to maintain, and waste materials can be added over time. Hot composting requires more maintenance and you must have all the materials ready before you begin, but the compost produced is far superior.

COLD COMPOSTING

Cold composting can be done in a bin or a heap. A bin makes composting quicker as it retains moisture and heat, but it is difficult to turn unless you invest in a compost tumbler. A heap is easier to turn but is exposed to the elements and the compost can suffer from being too wet or too dry.

For beginners and those with small spaces, I recommend using a compost bin. There are many options available, from recycled plastic bins and wooden beehive models to stackable boxes – the choice is entirely up to you. Your local authority may also sell compost bins at a discounted rate. Alternatively, you can make your own composter from a plastic storage box or a dustbin. Search online for DIY video tutorials. If you have a larger garden, then a compost heap is a good option. Fence it in to keep the pile neat and easier to turn. Fencing can be made from old pallets or steel mesh secured to a wooden frame. If you prefer a readymade option, at garden centres you can find timber boxes that slot together.

Site your bin or heap on a level, well-drained spot and in shade. Add a shovel of soil from your garden to the bottom of the bin/heap to encourage worm activity. Add materials in thin layers, aiming for 50% green waste and 50% brown waste (see pages 52–53). Greens are rich in nitrogen and include food scraps, grass clippings and weeds, whereas browns are carbon-rich materials and include newspaper, cardboard and dried leaves.

Once a month stir and turn the compost heap with a fork or shovel, mixing in any newly added organic matter. Turning compost helps to aerate, remove excess moisture and stimulate essential microbes.

Cold compost takes around six months to two years to be fully mature and ready to use.

HOT COMPOSTING

Hot composting is a quicker way of making compost than that of the compost bin/heap. It can yield compost that is ready to use in around three to six weeks, but it takes some skill to get it right. You need a larger sized bin or heap positioned in a sunny spot. All materials need to be ready as you can't add them later. The mixture is two parts brown materials to one part green materials (see pages 52–53). Make sure materials have been shredded or chopped to speed up the process.

Start by adding some twigs or branches to the bottom of the pile to help with airflow. Follow this by adding more brown materials. Water this mixture until it resembles a wrung-out sponge. Next, add some green materials and water if dry. Repeat each layer until you have a tall pile.

Leave the pile for three to four days. Turn the pile every two to three days. Cover with an old carpet or tarp to keep out rain in between turning. When the compost is turned, you may notice heat or steam coming from it. Hot composting requires a temperature of 55–63°C (130–145°F) to work effectively. To get an accurate reading, purchase a compost thermometer. Alternatively, place a metal rod or bar into the centre of the compost pile, pushing it down as far as it will go. When you remove the rod it should feel hot to the touch. If it does not feel hot, add some more green materials to restart the process.

After just three weeks your compost should be ready to use in the garden.

What to compost

Many people know that food scraps, grass clippings and newspaper are compostable, but there are lots of other things around the home and garden that can be added to the compost bin.

Here is a list of everyday items that you can add to the compost bin/heap, as well as some items that you may not have considered compostable.

Always shred or chop every item before you add it to the compost as this helps to accelerate decomposition. Don't add any cooked foods, meat or dairy as these can smell and attract vermin.

Greens

FROM OUTDOORS
» Grass clippings
» Weeds (without seeds)
» Annual plants and perennials
» Seaweed

FROM INDOORS
» Vegetable and fruit scraps
» Banana peel
» *Avocados:* separate the skin from the flesh and the pit, then cut the skin into strips before adding to the compost
» *Garlic:* can re-sprout if left whole, so chop into small pieces before composting
» Coffee grounds
» Tea leaves/bags (plastic-free only)
» Eggshells
» Houseplant cuttings/trimmings
» Flower bouquets
» Holiday greenery – evergreen wreaths, garlands etc

Browns

FROM OUTDOORS

» Straw/hay
» Dried leaves
» Pine needles/pinecones
» Nut shells (broken into tiny pieces, but avoid using walnut shells as these are toxic to plants)

FROM INDOORS

» *Paper:* newspaper, printer paper, coffee filters, napkins, cupcake/muffin wrappers, wrapping paper, junk mail
» *Cardboard:* cereal boxes, takeaway pizza boxes, egg cartons
» *Compostable packaging:* this is made from plants and includes sandwich wrappers, produce bags, takeaway coffee cups, straws, etc
» Cotton, linen, hemp and wool fabric
» Dried flowers
» Twine/jute
» Bamboo toothbrush handles – remove nylon bristles
» Cotton buds (cotton swabs), made from paper

» Toilet roll inners
» Dust or hair from sweeping/vacuuming
» Pencil shavings
» Plant-based/reusable sponges for cleaning/dishwashing
» Tampico fibres from wooden dishwashing brushes
» Natural rubber gloves
» Used matches
» Wood ashes
» Nail clippings
» Beard trimmings

Make your own indoor worm farm

Making your own worm farm couldn't be easier. A simple box made from wood or plastic is ideal for this purpose, but containers such as coffee cans, plastic tubs for chocolates (the family-sized ones sold at Christmas) or an old dishwashing bowl work well. If you use a container that doesn't have a lid, repurpose another item from your home. You will also need a second container to collect any excess liquid (see page 40 for how to make vermicompost tea).

Worms need to breathe so it's vital that you add air holes in your chosen container as well as a few holes in the bottom for drainage. Choose a location that is dry and dark for storing your box as worms don't like bright conditions – a kitchen cabinet or cupboard under the stairs is perfect.

YOU WILL NEED

» 2 plastic storage boxes or repurposed containers that fit one inside the other (one with a lid)

» Electric drill with a 12mm (½in) drill bit

» Spray bottle of water

» Worm bedding material – a mix of compost and shredded newspaper (you can also use coconut coir)

» Trowel

» Food (see page 49)

» Compost worms (Tiger/Red Wrigglers – see page 49)

1. Drill holes in equidistant rows in the container lid.

2. Drill multiple drainage holes in the base of the inner container. Flip the container onto its side. Near the top and under the lip, drill one set of holes the entire length of the container, leaving equal spaces between each hole. Repeat this again 2.5cm (1in) below your first row of holes

3. Repeat around each side of your inner container until all sides have holes. Insert into the outer container.

4. Spray the bedding with water until it feels like a damp sponge, then add to the container in an even depth of 5cm (2in). Use the trowel to dig a small trench in one corner, pour food into the trench, then cover with bedding.

5. Tip out the worms and any bedding they arrived in onto the layer of bedding. Add the lid and place in the chosen location.

6. Feed weekly, digging a small trench in one corner before adding food, then cover over with bedding. Increase the amount of food you provide as your worm population increases.

7. Check weekly for leachate (liquid run-off) in the lower container. Remove the inner container. Drain any leachate into a bucket or watering can, then replace the inner container. Dilute the leachate 1:10 with water to feed indoor and outdoor plants.

8. It will take a few months for the worm castings to be ready for garden use. Finished castings resemble good, rich soil. A week before harvesting, add food scraps to one corner of the bin. The worms should disperse to that area, leaving the remainder of the bin free of worms. Remove any castings with a trowel. Use as organic fertilizer or make vermicompost tea (see page 40).

Seeds

Seeds

Sowing seeds is a calming and nurturing activity, one that allows us to reconnect with nature and be fully present. I sow all sorts of seeds in my tiny backyard, from tomatoes and runner (string) beans to wildflowers and brightly coloured cosmos. I love the daily rhythm of checking on their progress and witnessing the first tender shoots emerge.

It is always fun choosing seeds to grow in the garden, yet it is all too easy to get carried away and buy and sow more than you need. Reducing food waste is paramount for the green gardener, so consider the amount of space you have before buying or sowing seeds. If it is a small area or you are growing in containers, concentrate on planting cut-and-come-again crops such as lettuce, spinach or rocket (arugula) rather than ones that take months to grow and only give you a small harvest.

The way foods are produced for supermarkets is also something to think about. If importing fruits and vegetables when out of season locally, transportation and refrigeration will all impact on the environment. Kale is shredded and bagged in non-recyclable plastic, cherry tomatoes and berries are placed in plastic punnets and broccoli heads are shrinkwrapped. These can all be grown at home, reducing the need for excess packaging.

Choose organic or heritage seeds rather than hybrid or non-organic ones (see page 60). Save seed from crops to grow next year and keep hold of packets of seeds that have passed their use-by date as they may still be viable (see page 73).

Even the most experienced gardener will have issues with seed germination, so on page 70 you'll find my hints and tips for resolving problems.

Choosing & starting seeds

Seeds are conventionally grown in intensive mono-culture farms with little consideration for the damage they are inflicting upon our planet. They are derived from plants that have been intensively sprayed with pesticides and herbicides. Conventional seeds may be fumigated or coated with agrichemicals to deter pests and prevent disease.

With this in mind, it is better to choose seeds that have been grown in a way that is less harmful to the environment. Organic seeds come from open-pollinated plants grown using methods that do not pollute water, air or the soil. These plants have not been sprayed with any synthetic fertilizers or pesticides and seeds are untreated. Heritage (heirloom) seeds come from plants that are open-pollinated and at least 50 years old. Heritage seeds for edible crops are well known for their exceptional flavour and

higher levels of nutrition. They are non-GMO, non-hybrid and untreated. You can save seed from both organic and heritage varieties.

There are two ways to sow seeds – indoors or direct sowing. Starting seeds indoors gives you a head start on the growing season, whereas with direct sowing, seeds are planted into the ground. Always read the backs of seed packets for dates regarding indoor and direct sowing.

TIPS FOR INDOOR SOWING

» *Air flow:* If air flow is poor around seedlings, the soil remains damp and diseases can take hold. Open a window on warmer days or use an oscillating fan to replicate a gentle breeze.

» *Light:* Seedlings need a lot of light to grow well. Place seed trays or pots on a sunny windowsill and remember to turn the containers regularly to prevent seedlings from becoming thin and leggy (see page 71). If low lighting is a problem, you may wish to install grow lights which can be easily purchased online.

» *Seed compost*: Ordinary potting compost is generally too rich and too coarse for seedlings to thrive. I like to make my own seed potting mix using a combination of homemade compost (see pages 46–53), leaf mould (see page 35) and sand. Mix equal quantities of sieved compost, leaf mould and sand in a bucket or large container. Cover with a lid and store in a cool, dry place.

If you do not have compost and leaf mould to make your own potting mix, buy seed composts from garden centres and organic gardening websites. Sustainable seed composts made with wool or biochar (see page 34) are good options for the green gardener.

» *Temperature*: Seeds need warmth for germination. The top of a refrigerator is a good option indoors, but a propagator or heat mat is a worthwhile investment if you plan on growing a lot of seeds. These maintain a constant temperature, speeding up germination and leading to a higher success rate.

continued »

» *Water:* Use a spray bottle to gently water seeds from above. Once germination occurs, water the seedlings from below (see page 67).

TIPS FOR DIRECT SOWING

» *Planting area:* Weed and dig over any vegetable beds. Level the surface with a rake and dampen the soil before sowing seeds.

» *Water:* Keep the soil moist until the seeds germinate, then water whenever the soil looks dry.

» *Temperature:* Different areas of the country warm up faster than others. I live in the north of England where temperatures are slower to rise than in the south, and I have found we are about two weeks behind in terms of ideal growing conditions. I always sow seeds two weeks after the dates given on the seed packet for this reason. Be mindful of your location when it comes to sowing seeds outdoors and always check the local weather report for frost warnings. Search online for your area's hardiness zone.

Containers for sowing seeds

There are many household items you can repurpose for sowing seeds.

SINGLE-USE PLASTIC CONTAINERS

Plastic fruit punnets, takeaway (takeout) containers and ready-meal trays are perfect as seed trays, and those with clear plastic lids can also be used as cloches. Small yogurt pots also make good seed starters for sunflowers, tomatoes, (bell) peppers and squash plants. Wash and punch drainage holes in the base of any plastic container before sowing. Fill your chosen containers with seed potting compost (see page 61) and lightly spray with water. Plant seeds as directed on the packet, spray again and place in a sunny spot.

COFFEE PODS

Millions of single-use coffee pods are sent to landfill as they are made of mixed materials and difficult to recycle, so use as seed starters. Compost any coffee grounds, wash the pod and leave to air dry. Poke a hole in the bottom with a sharp needle or awl. Fill with seed potting compost and lightly spray with water. Plant seeds as directed on the packet, spray again and place in a sunny spot. Once seedlings are ready for transplanting, gently squeeze the base of the pod to release the compost. Good for herb seedlings like basil, chives and coriander (cilantro).

TOILET ROLL INNERS (BIODEGRADABLE)

Squash a toilet roll inner flat with your hand, then fold it in half lengthways. Open out and at one end, cut by 2.5cm (1in) inwards along each crease in the cardboard to make four flaps. Moving clockwise, fold each flap downwards to form the base of the pot. Fill with seed potting compost and lightly spray with water. Plant seeds as directed, lightly spray with water once more and place

continued »

in a sunny spot. As the cardboard inner will decompose naturally, seedlings can be transplanted directly into the ground or container. Useful for all seeds.

EGGSHELLS (BIODEGRADABLE)

Perfect for herbs and salad seeds. Eggshells can be planted directly into the soil and as they decompose, they provide plants with additional nutrients. Gently tap the top of the egg and, peel away the shell, then pour the egg into a bowl and set aside. Wash the empty shell in clean water and leave to air dry. Carefully pierce a small hole in the bottom of the shell for drainage (a needle or an awl works well for this). Return the empty eggshells to their egg carton and gently spray them with water. Fill each eggshell with seed potting compost and plant seeds as directed on the packet. Lightly spray with water again and place in a sunny spot. Once seedlings are ready to be transplanted, gently crack the eggshell and plant directly into the ground or container.

NEWSPAPER (BIODEGRADABLE)

Lay two sheets of newspaper on a flat surface. Cut the newspaper into thirds lengthways. Lay a tin can (approx. 400g/14oz size) on the end of the newspaper closest to you and leave 2.5cm (1in) of newspaper above the top of the can. Think of it as gift wrapping a bottle – you need to be able to fold down the extra paper to form ends. Roll the can up the length of the newspaper until completely wrapped. Be careful not to make the paper too tight as this will make the can difficult to remove. Fold the edges of the paper in tightly to form the bottom of the pot. Sit upright

and then slide the can out. Fill the newspaper pot with seed potting compost and lightly spray with water. Plant seeds as directed, lightly spray with water and place in a sunny spot. As the newspaper will decompose naturally, seedlings can be transplanted directly into the ground or container. Useful for all seeds.

Make your own seed tape

Seed tape is used by gardeners to ensure that seeds are spread out evenly and waste is kept to a minimum. Making your own biodegradable seed tape will mean you can choose exactly which seeds you want to sow. Refer to seed packets for detailed instructions and ideal spacings.

YOU WILL NEED
- » 3 teaspoons plain (all-purpose) flour
- » 1 teaspoon water
- » Bowl
- » Unbleached toilet paper
- » Small artist's paintbrush
- » Seeds

1. Mix the flour and water in the bowl to form a paste.

2. Unroll the toilet paper to an appropriate length, then fold down the middle lengthways. Open out the paper along the crease. Towards the middle of one side, using the paintbrush lightly dab the paper with the flour paste at regular intervals.

3. Place a seed on each paste dot. Fold over the paper to seal the tape. Leave to dry, then plant following seed packet instructions.

Sowing seeds

Below is a general guide for sowing seeds indoors. Please refer to The Edible Garden chapter (pages 92–115) for advice on individual crops and always check seed packets for any specific instructions.

YOU WILL NEED
- » Seed potting compost (see page 61)
- » Bucket
- » Water
- » Trowel
- » Containers (for small seeds use a seed tray and for larger seeds use individual pots)
- » Seeds
- » Spray bottle filled with water
- » Plant labels (see page 23)

1. Place the seed potting compost in the bucket. Add a little water and mix it into the compost with the trowel. Moistening the compost before sowing seeds helps the mix to retain water and kickstart germination.

2. Fill your chosen container with moist seed compost. For small seeds, sow thinly and cover lightly with potting compost. For large seeds, drop two or three seeds in each container and cover lightly with potting compost.

3. Lightly spray the surface with water and place on a sunny windowsill. Label each container with the plant name and the date sown.

4. Check moisture levels daily and continue to spray water from above.

5. Once the seeds germinate, switch to bottom watering as this allows plants to develop stronger roots and helps to prevent disease. Place seed trays or pots on a tray or in an old basin. Add around 5cm (2in) of water to the tray/basin. Once the compost has turned a dark brown colour, the plant has absorbed enough water.

Transplanting, fertilizing & hardening off

TRANSPLANTING SEEDLINGS

Seedlings produce a first set of leaves called cotyledons. The subsequent sets that follow are known as true or adult leaves, and these will resemble and even smell like the vegetable variety you are growing. Once the seedlings have developed two or three sets of these adult leaves and you can see roots beginning to poke out the bottom of the container, it is time to transplant them into larger pots.

Choose containers that are approximately 25–30% larger than the original container. Larger sized yogurt pots are ideal for tomatoes, (bell) peppers and sunflowers. You could also re-use plant pots from garden centres or purchase biodegradable pots.

Water seedlings before transplanting. Fill containers about three-quarters full using a good quality potting mix (see page 12). Make a hole in the compost with your fingers big enough to fit the root ball. Carefully turn the seedling container upside down and squeeze the bottom to release the compost. Gently grasp the stem of the plant and pull the seedling loose from the container. Place the seedling into the hole and backfill with potting mix.

FERTILIZING SEEDLINGS

Seedlings will benefit from some extra nutrients after transplanting. A seaweed-based fertilizer or vermicompost tea diluted to a quarter of the regular strength is best for young seedlings (see pages 40 and 41). Fertilize one week after transplanting and every two weeks after that.

continued »

HARDENING OFF

Seedlings grown indoors or in a greenhouse need to be acclimatized to outdoor conditions before transplanting. Exposure to wind, rain and sun helps them grow stronger and become more able to survive the transplanting process. Hardening off takes approximately two weeks. On week one, place plants in a sheltered, shady position during the day, returning them indoors or to the greenhouse overnight. On week two, leave them out all day and night, but be mindful of late frosts. Plant on a cloudy day and water well.

Problem solving

It is always disappointing when seeds do not germinate or when the seedlings you have lovingly nurtured wither and die. Most problems are caused by overwatering, but timing, temperature and light conditions also play their part.

To reduce the risk of disease passing to vulnerable seedlings, use a seed potting mix (see page 61) rather than reusing old potting compost or soil from the garden. Wash any seed trays or pots in warm, soapy water and leave to air dry before every use.

Problems are less likely to occur when seedlings are watered from below. Fill a shallow tray or dish with water and place the seed tray or pot in the water. Once the soil looks moist on the surface, remove the tray or pot from the water and leave to drain freely. Alternatively, use a fine misting spray bottle to water seedlings gently from above.

COMMON PROBLEMS WITH SEEDLINGS

» *Damping off:* A general term used to describe the sudden death of new seedlings. Damping off diseases are caused by fungi in soil or water. It is usually seen in seedlings growing indoors or in a greenhouse and it thrives in wet conditions. Seedlings may wither or collapse and there is often white mould present on the surface of the soil. At the first sign of damping off, remove any affected seedlings and treat the remaining ones with an anti-fungal chamomile tea: add 1 chamomile teabag to 2 cups of boiling water. Leave to cool, then dilute with water until the tea turns a pale yellow colour. Pour the tea into a spray bottle and apply every two days until the seedlings are established.

» *Fungus gnats:* These lay eggs in the soil and once hatched, the larvae feast on the roots of seedlings. Apply a light dusting of ground cinnamon across the top of the soil in seed trays and pots and repeat monthly.

» *Leggy seedlings:* If your seedlings are tall, thin and not growing straight up, this is due to the light source being too far away or overcrowding as plants compete for light and resources. Move seedlings to a sunnier location or use a grow light system and be sure to thin out seedlings regularly.

» *Mouldy potting compost:* Green or fuzzy white mould on the top of seed compost means the soil is too wet. Scrape any mould off the surface and water the trays or pots from below rather than above.

Saving, swapping & using up seeds

SAVING

One of the cheapest and greenest ways to propagate your garden is to save seeds and use them the following year. Peas and beans, (bell) peppers and tomatoes are all easy to save seed from and they will produce a good plant the next year.

» *Peas and beans:* Leave pods to dry on the plant or harvest the plant and hang the pods upside down to dry indoors. Pop the peas or bean seeds out by hand and leave to dry on a piece of paper towel or reusable bamboo cloth. Remove any seeds with holes or discolouration. Transfer to an envelope, seal and label. Place in an airtight container.

» *Peppers:* Let the fruits fully ripen on the plant before picking. Cut in half and remove the seeds. Place the seeds on a piece of paper towel or reusable bamboo cloth and leave to dry. Once completely dry, transfer to an envelope, seal and label. Place in an airtight container.

» *Tomatoes:* Leave fruits to ripen on the vine. Pick, then cut in half and scoop out the seeds. Place in

a sieve and run under cold water to remove the sticky gel. Spread seeds on a piece of paper towel or reusable bamboo cloth and leave to dry. Once dry, transfer the seeds to an envelope, seal and label. Place in an airtight container.

» *Flowers:* Seeds from annuals such as calendula, cosmos, larkspur, nasturtiums, poppies and sunflowers can also be collected and saved to use the following year. Once flowers fade and turn brown, harvest the seed heads or pods that are left behind. Do this on a sunny, dry day as wet conditions can cause seeds to rot. Remove the head or pod with a pair of secateurs (pruning shears) and place in a brown-paper envelope or bag. Label each envelope or bag. After a few days, gently shake each seed head or pod, catching the falling seed in the envelope or bag. Store in a cool, dark place.

SWAPPING

Look online for seed swapping events near you. Most are run by local organic vegetable growing organizations and are usually held in early spring. If you do not have any seeds to swap, you can still attend the event to purchase seeds, chat to other gardeners and pick up a few hints and tips, too. Alternatively, search for online community seed exchange projects that allow growers to swap seeds by mail.

OUT-OF-DATE SEEDS

If you have flower and vegetable seeds leftover from the previous year or past their use-by date, do not throw them away as they may still be viable. Try this test to see if they will germinate. Place a few seeds between two pieces of damp paper towel. Fold the paper over and place in a sandwich/freezer bag. Place the bag somewhere warm – the top of the refrigerator or a sunny windowsill is ideal. Re-moisten the paper daily using a spray bottle if it becomes dry. After 8-10 days, open the paper to see if the seeds have germinated. If they have, they are ready to be planted.

Biodiversity

Biodiversity

Our ecosystem relies upon pollinators. These insects travel from flower to flower, transferring pollen to the female species of a plant, leading to fertilization and successful crop production. Almost all fruit and grain crops and more than half of the world's oils and fats are derived from animal-pollinated plants. The most common pollinators are bees and butterflies, but bats, moths, wasps and beetles also play a significant role. Sadly, many pollinating species are in decline, due to chemical pesticides, loss of habitat, disease and climate change.

However, there are easy ways to tempt pollinating insects into your garden. Plant native wildflowers and grow a mix of different plants to provide food and shelter for pollinating insects. If space is limited, a container filled with cosmos or a window box of thyme will provide a bountiful floral feast for pollinators. Provide shelter in the form of a bug hotel or leave a small patch of grass to grow longer and provide a home for beetles.

Learn to love your weeds! Nettles and dandelions are an important part of the ecosystem. They offer nourishment for pollinating insects and provide shelter and a place to breed. Leave a small area in the garden for weeds to flourish, or mow the lawn less often in spring when dandelions are most prolific.

Pests are a problem for every gardener. Aphids, slugs, snails and caterpillars can destroy crops and plants. Encouraging natural predators into the garden can help – hedgehogs devour slugs and snails, ladybirds (ladybugs) feast on aphids and birds love a juicy caterpillar. Nocturnal pests like codling moths provide a welcome supper for bats, and amphibians enjoy slugs, snails and flies.

Plants for bees & other pollinators

Bees and pollinating insects need three things to thrive: food, water and shelter. Whether you garden on a balcony, patio or allotment (community garden), it is easy to provide these insects with all three.

Pollinators feed from plants rich in nectar and pollen. Nectar contains sugar for energy; pollen has protein and oils. Bees and pollinators are active all year, so it is crucial for them to feed on plants rich in nectar and pollen throughout the changing seasons. If you can, include one or two plants for bees and pollinators each season (see right). A small pot of thyme on a sunny windowsill, an autumnal honeysuckle climbing up a trellis or a tiny clump of snowdrops pushing through the snow in winter are all great plants to grow for foraging pollinators.

Bees and insects need water to drink, and honeybees also use water to cool the hive in hot weather. Fill a shallow bowl or tray with water (preferably rainwater) and add a couple of large stones or pebbles for pollinators to rest safely on as they drink.

Provide shelter in the form of bee and insect boxes – it is simple to make your own from repurposed items (see pages 82–83 for how to make a bug hotel).

In periods of warm weather, you may find solitary bees exhausted and struggling to move. Feed them a quick energy mix of sugar water. Mix 2 tablespoons of sugar with 1 tablespoon of water. Pour a little next to the bee and watch it drink. After a short period of time the bee should be well enough to fly.

Seasonal plants for bees
& other pollinators

SPRING

» *Bulbs*: Bluebell, crocus, grape hyacinth, snake's head fritillary

» *Edible:* Blueberry, redcurrant, rosemary

» *Herbaceous perennials:* Ajuga, armeria (thrift), aubretia, geranium, primula, wallflower

» *Shrubs:* Berberis darwinii, ceanothus, daphne, rhododendron, ribes, skimmia japonica

» *Trees:* Acer, almond, amelanchier, apple, cherry, holly, peach, pear, pussy willow, tulip tree, willow

SUMMER

» *Annuals:* Borage, calendula, cosmos, nicotiana, nigella

» *Climbers:* Hydrangea, jasmine, nasturtium, passionflower

» *Edible:* Blackberry, broad (fava) beans, chives, courgettes (zucchini), fennel, marjoram, mint, raspberry, rosemary, runner (string) beans, sage, squash, strawberry, thyme

continued »

» *Herbaceous perennials:* Achillea, aquilegia, astrantia, dahlia, echinacea, echinops, foxglove (biennial), hollyhock, monarda, peony, perovskia, rudbeckia, scabious, verbena bonariensis

» *Shrubs:* Buddleja, escallonia, fuchsia, lavender

» *Trees:* Hawthorn, laurel

AUTUMN (FALL)

» *Climbers:* Honeysuckle

» *Herbaceous perennials:* Anemone, dahlia, echinacea, michaelmas daisy, rudbeckia, salvia, sedum, sunflower

» *Shrubs:* Abelia, arbutus (strawberry tree), hebe

WINTER

» *Bulbs:* Crocus, snowdrop

» *Climbers:* Clematis (evergreen varieties), ivy, honeysuckle (winter flowering varieties)

» *Herbaceous perennials:* Heather (winter-flowering varieties), hellebore, primrose, winter aconite

» *Shrubs:* Fatsia japonica, mahonia, sarcococca

If you are growing vegetables, leave a few plants to go to seed. The flowers of broccoli, kale and leek are a wonderful source of nectar for bees and other pollinators.

Growing cover crops as a green manure (see page 37) also benefits pollinating insects. Cover crops produce huge amounts of nectar and pollen and are particularly useful in early spring when other food sources may be scarce. Pollinating cover crops need to be allowed to bloom before digging them into the soil. Clover (red and white), comfrey, field beans, mustard and phacelia are all good options to grow for pollinating insects.

If space is limited, grow some nectar-rich herbs in a window box or container for pollinators. Chives, oregano, marjoram, rosemary, sage and thyme are all good options. Don't forget to add a few native wildflowers to your garden (see page 84).

Bug hotels

Building a bug hotel in the autumn provides beneficial insects and amphibians with a place to shelter, hide from predators or hibernate in during winter.

The best bug hotels have lots of small spaces in different shapes and sizes. They should be made of materials that blend in with their natural surroundings and ideally located in a shaded area near a tree or hedge.

On the following pages are three different ways to make a bug hotel – one utilizing an empty wall basket, another using a plastic bottle and, for larger gardens, one made from recycled pallets. Whichever bug hotel you choose to make, all will benefit from the addition of natural materials such as bark, twigs, pinecones, leaves, moss, bamboo canes, shells, stones, flowers or nutshells.

Frogs and toads like cool, damp conditions, so add a few stones or broken bits of tiles/terracotta pots if you want to attract them. Solitary bees will appreciate bamboo canes or old logs with holes of various sizes drilled through them. Add dried leaves, twigs and bark for ladybirds (ladybugs), beetles, spiders and woodlice.

Build your own bug hotel

Bug hotels can be large 'mansions' with many rooms, or small and compact. Both are appreciated by bugs and are equally effective. Here are three ideas that you can adapt to suit your outside space.

Upcycled pallet bug hotel

For larger gardens or allotments (community gardens). Good for attracting frogs and toads, as well as bugs.

YOU WILL NEED

» 3–5 wooden pallets
» Twine/cable grips
» Bricks, old tiles, broken terracotta pots and/or roofing slates
» A selection of natural materials
» Roofing materials, such as roofing slates, tiles or turf

1. Layer each pallet, one on top of the other. If they are not stable, tie them together securely with twine or cable grips.

2. Start with the bottom layer, adding bricks, old tiles, broken pots or roofing slates into the gaps.

3. Fill the gaps in each layer with more natural materials.

4. Finally, add a roof using roofing tiles or pieces of turf.

Wall basket bug hotel

A good way to utilize wall troughs and hanging baskets when summer displays have finished.

YOU WILL NEED

» Wall trough or hanging basket
» Pinecones
» Dried leaves
» Moss

1. Remove any liners from the trough or basket.

2. Fill the base of the trough or basket with a layer of pinecones.

3. Add a layer of dried leaves, then a layer of moss. Top with pinecones until the trough or basket is filled.

Plastic bottle bug hotel

Perfect for small spaces including balcony gardens to attract ladybirds (ladybugs) and other minibeasts.

YOU WILL NEED

» Large plastic water bottle or milk carton
» Scissors
» Natural materials

1. Carefully cut out a rectangular shape from one side of the bottle or carton.

2. Starting from the bottom, create layers from different natural materials (pinecones, bamboo canes cut to size, moss and small twigs are perfect).

3. Once it is filled from bottom to top, place in a warm and sheltered area.

Wildflowers

In the UK wildflowers are being lost at a rapid rate and one in five are under threat. Only 3% of flower meadows that existed in the 1930s remain*. This is happening worldwide and when wildflowers perish, vital food sources and habitats are destroyed.

The easiest way to help is to grow wildflowers in the garden. Start with this one change: stop mowing the lawn. Leaving a patch of grass to grow allows nature to take over and wildflowers to flourish. Try leaving a small patch around a tree trunk or a thin strip across the lawn. If long grass doesn't appeal, devote a patch of the garden to wildflowers: a meadow, a flower bed or border, or even in a container.

* The Wildflower Garden, part of Plantlife, a charity working to save threatened plants, www.plantlife.org.uk and www.plantlife. love-wildflowers.org.uk.

There are two types of wildflower meadow – annual or perennial. Annual meadows provide flowers throughout summer and autumn (fall), offering a rich source of nectar and pollen for insects. Usually a bright mix of poppies, cornflowers and marigolds, they are easy to recreate in a bed or border, but will need to be re-sown every year. Perennial meadows combine annual and perennial wildflowers as well as essential grasses. As well as nectar and pollen, these offer butterflies and moths a safe place to breed. Perennial meadows like poor soil and are better suited to large gardens or allotments (community gardens). They can be labour intensive.

Look for nurseries specializing in native wildflowers and check that the seeds or plants have local provenance. You can sow seeds, use plug plants or lay prepared wildflower turf.

Make a wildflower container garden

I like to fill patio pots and window boxes with wildflowers, and I use a seed mix that is specially developed for container gardening. Wildflowers need space to grow, so choose the largest container you can. A large window box, trough or ceramic pot work well. I also like to scour vintage and salvage fairs for old tin baths and wooden crates that I can repurpose for growing flowers and edibles.

YOU WILL NEED

» Large container with drainage holes
» Stones/broken crockery or terracotta pots
» Potting compost
» Wildflower seeds

1. Line the bottom of your container with a few pieces of broken terracotta pot or stones.

2. Fill the container with potting compost and sprinkle the wildflower seeds over the surface of the compost.

3. Lightly cover the seeds with a little more potting compost and water well.

4. Water your wildflower seeds twice a week, keeping the potting compost damp while the seeds germinate. You may need to water more frequently in hot weather.

Wildlife gardening

My little garden is close to a busy road. Traffic hums continuously in the background, the city pulsating with life, yet I feel completely removed from it. Here, in this tiny space, nature is abundant, my garden alive with foraging bees and butterflies. Birds tweet and sing, hopping from branch to branch before alighting to take a quick wash in the bird bath. At dusk, the garden is visited by bats and moths, as well as the occasional hedgehog.

Wildlife offers fantastic benefits to the organic gardener. It helps maintain a natural balance, from pollinating crops and flowers to keeping pests under control.

Wildlife needs three things: food, water and shelter. Food comes from plants or other insects and water from bird baths and ponds. Shelter can be provided by leaving areas of grass or plants to grow rather than be cut or by installing ready-made boxes (see pages 82–83 for how to make a bug hotel).

WAYS TO ATTRACT WILDLIFE TO THE GARDEN

» *Grow a hedge:* Hedges offer shelter for breeding and nesting, plus food in the form of berries, fruits or seeds. Choose informal, natural hedging such as beech, blackthorn, pyracantha or hawthorn. Rambling plants, such as honeysuckle and wild rose, growing through hedging offer extra foraging opportunities for birds and insects. Hedges and trees have both been proven to significantly reduce air pollution, too (see page 24).

» *Make a log pile:* A log pile offers shelter for insects and will attract birds, hedgehogs and frogs searching for food. Scatter logs (ideally from native trees) under a hedge or at the back of a flower border.

» *Amphibians:* Frogs, toads and newts are the natural predators of slugs and snails, as well as many crawling and flying invertebrates. Adding a sunken garden pond or one made from a container provides a home for amphibians and offers hedgehogs, bees and birds a place to drink. Ponds can also attract bats, damselflies, dragonflies and other insects, as well as providing a breeding area for frogs, newts and amphibians. You can make a pond to help wildlife from an old Belfast sink, oak barrel, dishwashing bowl or large planter. Include a few oxygenating plants in the pond too (specialist aquatic plant nurseries can advise on native species and help you choose the right size of plant for your pond).

Search for video tutorials online for how to build a container garden pond. Be mindful of children and pets around ponds and water features and add a fence if necessary. Alternatively, add netting that is protective yet still accessible for wildlife.

» *Bats:* These are natural predators for midges, flies, moths and mosquitoes. Grow night-scented plants and pale flowers such as dog rose, evening primrose, honeysuckle, night-scented stocks and nicotiana, which attract moths and therefore their predators, bats. Ponds are also useful as they provide drinking water and places to forage for insects. Avoid using artificial lighting in spring and summer as light beams can disorientate flying bats.

continued »

» *Birds:* These are natural predators for slugs, snails, aphids and caterpillars. Install a bird feeding station and a bird bath. Clean bird tables and bird baths regularly. Feed supplementary foods such as bird seed, suet balls or mealworms, and be mindful that birds have different nutritional requirements throughout the year.

» *Pinecone bird feeder:* Gather large pinecones from the woodland or forest floor (do not pick them from trees). If they are tightly closed, pop them in the oven for ten minutes on a low temperature to open the branches. Attach a length of jute string or twine to the tip of the pinecone. Spread peanut butter all over the pinecone. Pour some birdseed onto a plate and then roll the pinecone in the birdseed until it is completely covered. Hang outside from a tree or near a window so you can watch the birds feasting.

» *Orange rind bird feeder:* Cut an orange in half and scoop out the fruit, saving it to eat later. Using a pair of scissors, make a small hole near the top of the rind and the

same on the opposite side. Cut a piece of jute string or twine and thread it through these two holes, tying the ends together to form a hanging loop. Fill the orange with bird seed and hang outdoors. To make landing and feeding easier, create a perch using skewers or small twigs. Push a skewer or twig into the rind to make a hole, then through the orange to make a hole on the other side. Repeat on the opposite side with the second skewer or twig to create a cross shape.

» *Eggshells:* Use up old eggshells by feeding them to the birds in the garden. Eggshells are full of calcium and a good source of nutrition for nesting birds. Wash and air-dry the eggshells before sterilizing them in the oven (place on a baking tray for 30 minutes at 130°C (260°F). Crush in a food processor or by hand and then mix with wild bird seed.

» *Let flowers turn to seed:* Globe thistle, perennial asters, sunflowers and teasels are wonderful for birds in the colder months.

» *In the autumn (fall):* Leave apples, berries, haws and rosehips for birds to forage before migration.

» *Butterflies:* All butterflies need a safe place to lay eggs and form chrysalides. They also need a food source for the hatched larvae and nectar-rich plants for adults to thrive. Allow some areas of the garden to go wild. Nettles, thistles, holly and ivy are all good places for butterfly larvae to feed and to provide shelter for overwintering caterpillars. Choose nectar-rich plants with red, yellow, pink, purple and orange flower blossoms as these are particularly appealing to butterflies. Create a butterfly feeding station using a metal or glass dish. Add rotting or overripe fruits cut into slices to the dish. Bananas, oranges, strawberries and apples are all loved by butterflies.

» *Hedgehogs:* These are natural predators of slugs and snails. Provide hedgehogs with access to the garden by creating tunnels or holes in fencing. Build a hedgehog house and situate it in a quiet, shady spot of the garden. Search for online video tutorials on how to build a hedgehog house or, alternatively, buy a readymade option from garden centres. Encourage hedgehogs to visit the garden by offering some food and a shallow dish of water (don't put out a saucer of milk as this can make them ill). Specialist hedgehog food can be purchased at pet stores or garden centres.

The Edible
Garden

The edible garden

There are so many positive reasons for growing your own food: less food miles, no synthetic pesticides and herbicides are used, and biodiversity is increased. But by far the best thing about growing your own food is how good it tastes.

First think about what you like to eat and how much spare time you have to work on a vegetable garden or the size of plot you have. Newcomers to vegetable growing often plant too many crops, resulting in an overcrowded and unproductive plot. It is unlikely that you will be able to grow all the crops your household needs, so continue to buy produce that is tricky to grow or needs a lot of space for minimum return. For example, asparagus crowns take four years to reach maturity and Brussels sprouts take months, but spinach, rocket (arugula) and salad leaves can be sown in succession and feed you all season long.

Growing your own also gives you the opportunity to try out different varieties of fruits and vegetables to those in the grocery store. It is also worth prioritizing fruits and vegetables that you would usually buy pre-packed, such as fruits in single-use plastic punnets or potatoes or leaves in plastic bags.

For a productive plot, most plants require 6–8 hours of direct sunshine during the day. However, some crops will do well in shady areas (see page 101). Some plants are more reliable indoors, like cucumbers, tomatoes and aubergines (eggplants); they can be grown outdoors but will need shelter from the wind.

If you produce too many seedlings for your space, consider donating them to community gardens, fund-raising events or school fairs. Donations of fresh produce to food banks are often appreciated.

Vegetables

All the vegetables listed here are easily grown from seed. However, if you are a beginner to vegetable gardening, I would not recommend growing everything from seed at first. Instead, try growing one or two things and supplement your veg patch or balcony garden with plug plants from a plant nursery. Good plants to start from seed are beans and peas, courgettes (zucchini), tomatoes and any of the salad crops or leafy greens.

To make direct sowing easier you can create your own seed tapes (see page 65) which reduce waste and help to space seeds evenly.

Most vegetables will not require fertilizing if the soil is rich in organic matter and you are watering well. However, if crops do show signs of flagging, a good feed with an organic liquid fertilizer (see pages 38–43) should

suffice. Aubergines (eggplant), chillies (chiles), (bell) peppers and tomatoes are the exceptions as they will need regular fertilizing throughout the growing season.

» *Aubergines:* Sow in individual pots indoors in early spring. Plant out in late spring/early summer. Plants may need protection from the cold and support in the form of staking. Apply a layer of mulch around the roots of plants (see page 34). Feed with an organic liquid fertilizer every two weeks during the summer months (see pages 38–43).

» *Beans and peas:* These have long roots so need to be sown in containers that provide a good depth of soil. Toilet roll inners or newspaper pots are ideal for this purpose (see pages 63–65), and they can be planted directly into the soil as they are biodegradable. If you are planning to grow beans and peas, sow them in the order they would be planted out: broad (fava) beans, peas, runner (string) beans and French beans. Harden off all bean and pea seedlings before planting outdoors (see page 70).

» *Broad beans:* Sow indoors in early spring. Push the bean into the potting compost to a depth of around 5cm (2in), then cover over with a little more compost. Place on a sunny windowsill and water well. Once the seedlings are around 8cm (3in) tall, they are ready to be hardened off before planting.

» *Peas:* Sow indoors in early spring, with one or two pea seeds per container. Once seedlings have reached a height of around 10cm (4in), they are ready to be hardened off before planting. Peas will require support once planted out.

» *Runner beans:* Sow indoors in mid spring. Push the bean into the potting compost to a depth of around 5cm (2in), then cover over with a little more compost. Place on a sunny windowsill and water well. Once the seedlings are around 10cm (4in) tall, they are ready to be hardened off before planting. Runner beans will require support once planted out.

» *Dwarf French beans:* Sow indoors in late spring. Push the bean into the potting compost to a depth of around 5cm (2in), then cover over with a little more compost. Place on a sunny windowsill and water well. Once the seedlings are around 10cm (4in) tall, they are ready to be hardened off before planting.

continued »

BRASSICA FAMILY

All members of the brassica family benefit from plenty of organic matter added to the soil before planting. Sow seeds indoors in early spring or direct in mid to late spring. Harden off indoor grown seedlings before planting. It is also a good preventative measure to fix a collar around seedlings after planting to help deter cabbage root fly (see page 163). Water well the day before planting.

» *Broccoli:* Once seedlings have reached a height of 10cm (4in) they can be planted out. Space plants 45cm (18in) apart. Harvest the large central head, leaving the side shoots to grow and be harvested a few weeks later.

» *Brussels sprouts:* Once seedlings have reached a height of 10cm (4in) they can be planted out. Space plants 30cm (12in) apart. Harvest from the bottom of the plant first, working your way upwards. Twist each sprout to remove.

» *Cabbage:* For summer varieties, sow in late summer for planting out in autumn (fall). For autumn varieties, sow in late winter for planting in early spring. Once seedlings have reached a height of 8cm (3in), they can be planted out. Space plants 45cm (18in) apart.

» *Cauliflower:* Once seedlings have reached a height of 8cm (3in) they are ready to be planted out. Harvest when the head has fully developed but before the curds (or florets) start to loosen. If cauliflower heads turn a little yellow in colour, they are still good to eat. The leaves have simply opened and exposed the heads to the sun.

» *Kale:* Sow indoors rather than direct. Once seedlings have reached a height of 10cm (4in) they can be planted out. Space plants 40cm (15in) apart. Harvest the outer leaves first, leaving the smaller, tender leaves to grow and cut later. Leave kale in veg beds or containers over winter and it will flower the following spring. Bees and other pollinating insects will find it irresistible.

CUCURBIT FAMILY

Courgettes (zucchini), pumpkins and squash will benefit from plenty of organic matter being added to the soil before planting.

» *Courgettes:* Sow indoors in individual pots in mid to late spring. Press a couple of seeds on their sides into the potting compost. Cover lightly with the compost and water well. Harden off seedlings before transplanting outdoors. Plant after last frosts. Courgettes need around 60cm (24in) between plants. Feed every two weeks with an organic liquid fertilizer. Courgettes can be extremely productive so do not sow too many or you will be eating courgettes every day for weeks!

» *Cucumbers:* Sow indoors in individual pots in early spring. Sow two seeds per pot about 3cm (1in) deep and water well. Cucumbers grow best in a greenhouse (glasshouse), but some varieties are suitable to grow outdoors. Transplant seedlings into larger containers and keep indoors until late spring/early summer as they are prone to collapse in cold weather. Plant outdoors in the ground or in a large container. Cucumbers climb vigorously and will need to be supported, so add bamboo canes or a trellis system. You will need to remove the side shoots regularly and pinch out the tops. Search online for video tutorials to help you with this.

» *Pumpkins and squash:* Sow indoors in individual pots in early to late spring. Press a seed on its side into the potting compost. Cover lightly with the compost and water well. Once seedlings have established, transplant outdoors. Check seed packets for correct planting distances as these vary depending on the pumpkin or squash variety. Water well.

ONION FAMILY

» *Garlic:* Buy from a garden centre or nursery rather than using grocery store bulbs, which may carry disease or be unsuitable for your climate. Add a little organic matter to the potting compost. Plant individual cloves pointed end upwards in large containers in late autumn or early winter.

continued »

Dig up bulbs when plants have turned yellow.

» *Onions:* These can be grown from seed but I prefer to grow them from sets (these are essentially part-grown onions and are easier and quicker to grow than seeds). Plant in mid-spring, spacing them 10cm (4in) apart and at a depth approximately twice the height of the bulb. Leave the tip just showing on the surface of the soil. Harvest when the leaves turn yellow.

» *Spring onions (scallions):* Direct sow from spring in rows 1cm (½in) deep. Thin seedlings until they are 2.5cm (1in) apart. Water well. Harvest the largest plants first.

» *(Bell) peppers and chillies (chiles):* Sow indoors in early spring into small individual pots. Germination rates are better when seeds are placed on a heat mat or in a propagator, but if you are growing on a sunny windowsill, pop a plastic lid over the top of any pots to keep the heat in the soil. Once seedlings emerge, take them off the heat mat or out of the propagator and remove any plastic lids. Harden off seedlings before transplanting into larger containers. Water well. Once the plants start to produce flowers, feed regularly throughout the growing season with an organic liquid fertilizer. Bring containers indoors on colder days or nights.

ROOT VEGETABLES

» *Beetroot (beets):* Soak seeds overnight in water for quicker germination. Direct sow in mid spring and in mid to late summer for harvesting in autumn (fall). Lightly cover seeds with soil and thin seedlings to at least 10cm (4in) apart.

» *Carrots:* Sow direct from mid spring to late summer in rows 2cm (¾in) deep and 15cm (6in) apart. Thin seedlings to at least 5cm (2in) apart. Water well in dry spells. Harvest once you see the orange top sitting just above the soil.

» *Radishes:* Sow direct from early spring to late summer. Sow thinly along the row at a depth of 1cm (½in) and 10cm (4in) apart. Water well. Harvest the largest roots

first, leaving the small ones
to grow bigger.

» *Turnips:* Sow direct from mid
spring to midsummer. Sow in
rows at a depth of 1cm (½in) and
the same apart. Thin seedlings
until they are eventually 15cm
(6in) apart for early crops, or for
turnips to harvest in autumn, thin
seedlings until they are 23cm
(9in) apart.

SALAD CROPS AND
LEAFY GREENS

» *Chard:* Sow direct in mid spring
and in late summer. Sow in rows
10cm (4in) apart and 2.5cm (1in)
deep. Thin seedlings to 30cm
(12in) apart. Water well. Harvest
the outer leaves first, leaving the
smaller, tender leaves to grow and
cut later.

» *Lettuce:* Sow indoors in early
spring and throughout the
growing season for a continuous
harvest. Direct sow in rows 1cm
(½in) deep and 10cm (4in) apart.
Water well. Harvest whole heads
or remove the outer leaves, leaving
the centre to grow bigger and be
cut later.

» *Rocket (arugula)*: Sow direct in
early spring in rows 1cm (½in)
deep and 10cm (4in) apart. Water
well. Harvest young leaves for
salads or add to pesto.

» *Spinach:* Sow direct in early
spring to late summer in rows 1cm
(½in) deep and 2.5cm (1in) apart.
Thin seedlings once they reach
a height of 2cm (¾in) to 25cm
(10in) apart. Water well. Harvest
the outer leaves first, leaving the
smaller, tender leaves to grow and
cut later.

Tomatoes

These fall into two different categories: determinate and indeterminate. Determinate tomatoes are known as 'bush' tomatoes and they grow to a height of around 1m (3ft) tall. Indeterminate tomatoes are known as 'cordon' tomatoes, which grow on a single stem and have side shoots that need to be removed. Cordon varieties will need supporting. You can also grow cherry and tumbling tomato plants that are suitable for hanging baskets and patio containers.

Sow indoors in early spring in small pots. Pop two or three seeds on the surface of the potting compost and lightly cover with a little more of the mix. Place on a sunny windowsill and water well. When the seedlings have grown large enough to handle it is time to transplant them into larger pots. Tomato plants have roots all the way up their stems that resemble tiny hairs and wherever the stem hits the soil the plant will develop roots. When transplanting tomato seedlings, plant the stem deeper in the container, right up to the first set of leaves. Harden off before planting outdoors (see page 70). Plant in large containers for greenhouses (glasshouses) and outdoor growing.

» *Bush tomatoes:* Water regularly and feed every two weeks with an organic liquid fertilizer (see pages 38–43).

» *Cordon tomatoes:* Add a bamboo cane to the container and tie the main stem to the support. Water regularly and feed every two weeks with an organic liquid fertilizer. Remove all side shoots. Once the plants have developed four or five trusses, cut off the very top of the plant.

What to grow in shady areas

FRUITS

» Alpine strawberries

» Blueberries

» Gooseberries

» Rhubarb

» Sour cherries

VEGETABLES

» Kale

» Lettuce

» Potatoes

» Radishes

» Rocket (arugula)

» Spinach

HERBS

» Chives

» Coriander (cilantro)

» Mint

» Parsley

Potatoes

Like most vegetables, potatoes like to grow in full sun and well-drained soil. They do well in the ground or raised beds and they are also good grown in containers (see opposite for how to grow in a compost bag).

There are three different types of potato to grow: first earlies, second earlies and maincrop. First and second earlies are commonly known as new potatoes and are ready to harvest in spring. Maincrop potatoes grow larger and are typically lifted in late summer or early autumn (fall) and stored over the winter.

Blight can be a problem for potatoes, so growing first and second earlies can be the wisest choice as they are harvested before the peak risk time for blight (see page 162). If you do choose to grow maincrop potatoes, consider what you will use them for. Some

varieties are good for baking and mashing, whilst others make incredible chips (fries).

All potatoes are grown from seed potatoes (tubers) and you can buy these from garden centres or plant nurseries. Although you can use potatoes from the grocery store, seed potatoes have been certified virus free and are more reliable for planting.

It is a good idea to prepare seed potatoes for planting by 'chitting'. This encourages new green shoots to form on the potato and can lead to higher yields. Place the seed potatoes in an empty egg carton, positioning them with the eyes pointing upwards. Keep them in a cool, light place and the potatoes are ready to plant when the shoots have grown to 2.5cm (1in). Chit early and maincrop potatoes at least six weeks prior to planting.

How to plant potatoes

First earlies are planted in early spring and are ready to harvest in 10–12 weeks. Second earlies are planted in mid-spring and are ready to harvest in 14 weeks. Maincrop potatoes are planted in mid to late spring and are ready to harvest in 6–8 weeks.

Dig a trench to grow a row of potatoes or a hole for individual plants. Plant the tubers to a depth of approximately 15cm (6in) and space them around 30cm (12in) apart along the row. Cover the tubers with the soil from the trench or hole.

After two to three weeks you should see some shoots starting to poke through the soil. Once the shoots reach a height of 15cm (6in) above the soil, use a rake to mound soil over the top of them. This is known as 'earthing up' and is used to encourage more tubers to grow and to protect them from light exposure, which can turn potatoes green. Water well in dry weather.

Early potatoes are ready to harvest once flowers appear, whereas maincrops can be harvested once the plants start to die back. Dig up with a fork, approaching from the side of the plant to avoid damaging the crop.

Store maincrop potatoes in hessian (burlap) or paper sacks and keep in a cool, dark place.

Using potting compost bags

Save the plastic sacks that potting compost is sold in as they are ideal for growing potatoes. Make several holes in the bottom of the bag for drainage purposes. Fill the bag halfway with potting compost, then roll down the sides of the bag until they are just above the surface of the soil. Bury two or three seed potatoes in the potting compost. Water two or three times a week and once green shoots start to appear cover them over with a little more compost, unrolling the bag to make it taller. Feed container-grown potatoes with an organic liquid fertilizer once a month. Once flowers are produced, first and second earlies are ready to harvest and when plants start to die back it is time to empty your containers and harvest the maincrops.

Fruit

Supermarkets sell fruits that have been specifically bred for their appearance and storage qualities, rather than for their taste. Rather worryingly, many have been treated with chemicals to stop them deteriorating in cold storage. Fruits grown at home not only taste incredible, they are also healthier and give you the opportunity to try unusual varieties you would never find in the grocery store.

All fruit bushes and trees prefer a sunny and well-drained soil to grow well. The fruits listed below can all be planted outdoors in the ground and most can also be grown successfully in containers (see pages 108–112).

» *Blueberries:* Plant blueberries in a sunny location in moist, well-drained acidic soil. Blueberries do not like an alkaline soil, so always check the pH level before planting (see page 32). They are also not keen on clay soils, so growing in containers can be more productive (see page 110). Blueberries are self-pollinating but produce more fruit when planted next to a different variety. Feed with an organic liquid fertilizer suitable for ericaceous plants. Fruits are ripe when they turn a dusty blue colour.

» *Redcurrants and whitecurrants:* These prefer to be planted in sunny sheltered spots but can tolerate partial shade. Fruit grown in sun will ripen quicker and taste sweeter. Buy established plants with three to five stems from a garden centre or nursery in early spring. Plant bushes in the ground at the same depth as they were in the container. Mulch well after planting and do not overwater. Most established bushes will not need watering unless it is exceptionally dry. Bushes will need pruning over winter.

» *Raspberries:* Plant raspberry canes in late autumn (fall) or early spring in rich soil with a little compost added to the soil. Raspberries are often planted in rows and plants should be spaced approximately 60cm (24in) apart. There are lots of different varieties of raspberry to grow and harvest, some fruit in summer whereas others fruit in autumn. Like strawberries, it is a good idea to grow both types if you have space as this gives you fruits over a longer period.

Both types of raspberry need to be pruned for the plant to produce well the following year. Prune summer-fruiting raspberries after fruiting and autumn-fruiting varieties in spring. Summer-fruiting raspberries fruit on the previous year's wood so the old canes need to be cut out, and the new green canes are left. Autumn-fruiting raspberry canes can be cut right back to ground level.

» *Rhubarb:* Plant dormant rhubarb crowns in late autumn or early spring in a hole twice as big as the crown. Add a little compost or a small pile of comfrey leaves to the bottom of the hole, then plant with the tip of the crown just poking out of the soil. Rhubarb likes an open position and a deep, rich soil. Feed several times throughout the growing season with an organic liquid fertilizer and keep well-watered. Do not harvest the stems in the first year but remove a few the next, and then from the third year onwards, harvest up to half the stems.

continued »

» *Strawberries:* You can plant bare rooted strawberries or choose established plants in pots. Bare root strawberries look like woody twigs on the top with lots of roots below. In amongst the woody top is the strawberry crown and you can usually see the pink-red flesh peeking out. Plant bare roots in autumn (fall) and overwinter indoors or in a greenhouse. To plant them, fill small containers (large yogurt pots work well) with a good-quality potting compost, make a hole in the middle of the compost and twist the roots slightly to fit in the hole. You may need to trim the bottom of the roots if they are too thick or long to fit in the pot. Make sure the pink-red tip of the crown sits above the surface of the compost, then backfill. Transplant outdoors in spring for summer harvests. For strawberries bought as potted plants in late spring, plant outdoors straight away.

Strawberry varieties can be classed as early, mid and late fruiting. It is a good idea to buy a few of each type so you have continuous fruiting throughout summer. Protect fruits with a layer of straw or grass cuttings and be mindful of birds. Remove runners in autumn and replant into separate pots. If runners are left to root in the bed, they will produce smaller fruit the following year.

Fruit trees

A fruit tree makes a wonderful addition to the edible garden. It not only provides you with a beautiful display of springtime blossom, it also offers a vital source of nectar for pollinating insects and rewards you with a bountiful harvest at the end of the growing season.

There are lots of different fruit trees available to grow at home, from apples, figs and pears to damsons, peaches, plums and apricots. No matter which type of tree you choose, the following information applies to them all.

Some fruit trees are self-fertile whilst others rely on cross pollination from nearby trees. If you live in an urban area, your garden is more than likely to be close to other fruit trees and you do not need to worry about pollination. However, if you live in a more rural area, it is better to plant two trees. Crab apples are good for pollinating apple trees as they have a genetic compatibility and produce a lot of flowers over a long period of time.

Many fruit trees are grafted onto a different rootstock to limit the growth of the tree and improve its resistance to disease. Rootstocks are typically dwarfing, semi-dwarfing or semi-vigorous. Dwarfing trees will be under 3m (10ft) in height, semi-dwarfing will be under 4m (13ft) in height and semi-vigorous will be under 5m (16ft). You can also find extreme dwarfing varieties which are for growing in containers.

A good idea for small gardens is to train a fruit tree. There are four ways to do this: cordon, espalier, fan or step-over. A cordon is grown as a single stem with the short fruiting spurs spaced evenly along it. An espalier tree has a central trunk with two or three arms trained horizontally either side. A fan-trained fruit tree has multiple branches and needs to be grown next to a wall or a fence. A step-over is a single tier espalier used to edge a bed, border or wall.

Plant fruit trees in a sunny, sheltered spot with good-quality, well-drained soil. Container-grown trees should be planted in early autumn or in spring. Plant bare-rooted trees in late autumn or over winter if the ground is not frozen. Dig a hole to the same depth as the roots or the container but a third wider. If your tree was purchased in a container, water it well and leave for at least an hour before removing it from the pot. If it is a bare-rooted tree, soak the roots in a bucket of water for at least two to three hours before planting. Place the tree in the hole and add a stake for support. Tie this in with some jute twine, making sure the tree does not rub against the stake. Backfill with soil and gently firm in. Water well after planting.

Growing edibles in containers & raised beds

Although I grow mostly cut flowers and perennials in my tiny backyard, I do always make some room for herbs and my favourite edible crops. I love planting beans, peas and salad leaves, as well as some rocket (arugula), spinach and a few salad potatoes. I grow everything in containers, some found at salvage fairs, others repurposed from old watering cans or fruit crates.

Large containers are best for growing edible crops. Bigger pots hold more potting compost which can stay moist longer and retain more nutrients. Size, shape, colour and the material a container is made from also affect the moisture level in the potting compost as well as the temperature. Round, square and rectangular containers can retain moisture better than those that are tapered. Lighter coloured pots reflect light and keep soil moist and the temperature cooler. Darker coloured pots absorb heat so dry out quicker.

» *Terracotta:* Commonly used in gardens as it is relatively inexpensive, and it ages beautifully. However, as it is porous, air and water pass through its walls and soil dries out quickly. Before planting, soak terracotta pots overnight in a bucket of water to hydrate the clay. This helps to prevent the soil from drying out and protects the pot from cracking during cold weather. If your pot is too large to soak in a bucket, fill it up with water and allow it to evaporate naturally before planting.

» *Glazed:* Usually made with terracotta but glazed on the outside of the pot. The glaze prevents air and water from passing through its walls and so is better for retaining moisture.

» *Fabric:* Available in a variety of sizes and styles, these are lightweight, durable and can be washed and put away for use next season.

» *Reuse/repurpose/upcycle:* Plastic plant pots, metal colanders, wine crates, old buckets, ceramic sinks, etc can all be used for container gardening. If you are repurposing or upcycling an item, make sure it is clean and has adequate holes for drainage.

Fill containers with a good-quality potting compost suitable for growing edible crops. Do not use soil from the garden as this can become compacted quickly and may contain weeds and pests. At the end of the growing season, tip the spent compost on to beds as a mulch.

The best crops to grow in containers are those that give you a regular harvest. Tomatoes, salad leaves, spinach, radishes and beetroot (beets) are all good options. Some vegetables such as cauliflowers and cabbages will grow in containers, but need a large pot per plant and will take a long time to grow. Most plants need around six hours of direct sunlight every day, although some crops will do well in partial shade (see page 101). Water your container vegetables regularly. During periods of warm weather, you may need to water containers twice a day. Feed with an organic liquid fertilizer (see pages 38–43) every two weeks.

continued »

Below you will find a list of fruits and vegetables that grow well in containers. Most can be sown direct, but others will need to be sown indoors and transplanted later (see pages 60–62 for indoor sowing information). Fruit trees can also be grown in containers (see page 107).

EDIBLE CROPS FOR CONTAINERS

» *Aubergine (eggplant):* Minimum container size: 20cm (8in) deep. Sow indoors and transplant seedlings. Grow in dark coloured containers to help retain heat. One plant per pot. May require staking.

» *Beetroot (beets):* Minimum container size: 25cm (10in) deep. Plant seed approximately 1cm (½in) deep in the container. Water well. Thin seedlings to approximately 7cm (3in) apart.

» *Blueberries:* Minimum container size: 45cm (18in) deep. One plant per pot. Blueberries need an acidic soil so the pots must be filled with an ericaceous potting compost. Plant two different varieties of blueberry next to one another as

they cross-pollinate, resulting in a more productive crop. Feed blueberries with an organic liquid fertilizer suitable for ericaceous plants.

» *Carrots:* Minimum container size: 30cm (12in) deep for regular carrots. For smaller, compact varieties 15cm (6in) deep. Plant seed approximately 1cm (½in) deep in the container. Water well. Thin seedlings to approximately 5cm (2in) apart.

» *Chard:* Minimum container size: 25cm (10in) deep. Direct sow approximately 1cm (½in) deep in the container and 2.5cm (1in) apart. Thin seedlings to 6cm (2½in) apart. Harvest the outer leaves first, leaving the smaller, tender leaves to grow and cut later.

» *Courgettes (zucchini):* Minimum container size: 30cm (12in) deep. Look for compact, patio varieties. Sow indoors and transplant seedlings. One plant per pot.

» *Figs:* Minimum container size: plant into a pot one size larger than the original. Buy as an established plant. Add some garden compost into the container along with the potting mix. Water well. Feed with liquid organic fertilizer every two to three weeks once fruits appear. Plants should be relocated to a shed or unheated greenhouse (glasshouse) over winter. Some pruning may be required – search online for video tutorials.

» *Green beans:* Minimum container size: 30cm (12in) deep. Dwarf varieties work best in containers. Direct sow beans 5cm (2in) deep and approximately 10cm (4in) apart. Climbing (pole) beans will require some support.

» *Kale:* Minimum container size: 30cm (12in) deep. Choose compact or dwarf varieties. Sow indoors and transplant seedlings. Harvest the outer leaves first, leaving the smaller, tender leaves to grow and cut later.

» *Lettuce:* Minimum container size: 15cm (6in) deep. Sow direct into container, cover lightly with potting compost and thin seedlings regularly. Plants should be spaced approximately 7cm (3in) apart Harvest the outer leaves first, leaving the smaller, tender leaves to grow and cut later.

» *Peas:* Minimum container size: 15cm (6in) deep. Choose short or dwarf varieties. Soak seeds for 24 hours prior to sowing to speed up germination. Sow indoors or direct sow.

» *(Bell) peppers:* Minimum container size: 30cm (12in) deep. Sow indoors and transplant seedlings. One plant per pot. May require staking.

» *Chilli (chile):* Minimum container size: 30cm (12in) deep. Sow indoors and transplant seedlings One plant per pot. May require staking. When the plant reaches a height of 15cm (6in), pinch the growing tip to help it bush out. Remove any flowers from seedlings before transplanting.

» *Radishes:* Minimum container size: 15cm (6in) deep. Direct sow. Plant seed approximately 1cm (½in) deep and the same distance apart in the container. Water well. Thin seedlings to approximately 5cm (2in) apart.

continued »

» *Raspberries:* Minimum container size: 30cm (12in) deep. Buy compact raspberry canes which are ideal for container growing. Make a hole for each raspberry cane in your container. The canes should be equally spaced in the pot. Place a cane in each hole and backfill with potting compost. Water well.

» *Rocket (arugula):* Minimum container size: 30cm (12in) deep. Direct sow. Plant seed on the surface of the soil, then lightly cover with potting compost. Water with a spray bottle. Keep the soil moist and thin seedlings to approximately 2.5cm (1in) apart.

» *Spinach:* Minimum container size: 15cm (6in) deep. Direct sow. Plant seed approximately 1cm (½in) deep in the container. Water well. Thin seedlings to approximately 5cm (2in) apart. Harvest the outer leaves first, leaving the smaller, tender leaves to grow and cut later.

» *Strawberries:* Minimum container size: 10cm (4in) deep. Follow instructions on pages 105–106 for how to grow strawberries.

» *Tomatoes:* Minimum container size: 30cm (12in) deep. Choose compact, bush varieties for containers. Sow indoors and transplant seedlings. One plant per pot. Tomatoes need moisture, so water well (see page 100). Mulching is also a good idea for container tomatoes to help retain moisture (see page 34).

RAISED BEDS

If you have the space in your garden, a raised bed is a great way to increase the number of crops you can grow. Raised beds are good for veggie growing as they offer improved drainage and an increased soil temperature, and they can be placed over areas that would be impossible to plant in.

First, decide how long and wide your raised bed should be. For small spaces, a square 1.2m x 1.2m (4 x 4ft) is a good size. The most common size is 1.2m x 2.4m (4 x 8ft) which allows you to reach the middle of the bed on either side to tend to your crops. A popular height is 30cm (12in), which provides sufficient drainage for most crops. However, if you are locating a raised bed in an area that is paved, it is better to increase the height to allow more root space.

You can purchase readymade raised beds from garden centres, but it is easy to make your own from reclaimed materials. Bricks, wood, pallets and logs are good options. Be mindful of pressure treated timber as the chemicals used in its manufacture can leach into the soil. Line raised beds with old compost sacks, newspaper or invest in a good-quality liner to help prevent this. Search for online video tutorials for how to make a raised bed.

Most plants need at least six hours of sunshine a day to grow successfully, so raised beds need to be in a sunny spot. Fill with a 50:50 mix of good-quality topsoil and homemade compost. Feed crops once a week with an organic liquid fertilizer (see pages 38–43) and add a layer of organic mulch at the end of the growing season (see page 34).

When you are planting up a raised bed, consider both light and wind. Large and climbing plants may block out a lot of light for smaller crops. Taller plants may need to be staked in windy conditions or fixed to a trellis.

Grow your own pizza garden

In late spring I like to put together a container filled with tomatoes and fresh herbs that I can later harvest and add to homemade pizzas. I have an old vintage enamel bowl I use for this purpose, but you can use any large size container to make your pizza garden.

If you are using a vintage container or repurposing a household item, wash it in warm, soapy water, then rinse and leave to air dry. Make sure that you provide adequate drainage holes by drilling several holes in the bottom of the container. Wooden crates or boxes need to be lined with an old compost bag or a bin liner before adding the potting compost. Again, make sure there is adequate drainage by cutting a few holes in the liner.

Use a good-quality potting compost suitable for growing vegetables. I like to grow basil and oregano as they are a classic combination for pizza, but you could add thyme, marjoram or rosemary if you prefer. Choose tumbling cherry tomatoes as they will cascade over the side of the container and will not require staking. I add some rocket (arugula) leaves for a peppery tang, which I add to the top of pizzas after cooking. A compact chilli (chile) plant would also be a nice addition if the container is large enough to accommodate it.

YOU WILL NEED

- » 2 tumbling cherry tomato plants (suitable for growing in containers)
- » 1 pot basil
- » 1 pot oregano
- » 1 packet rocket (arugula) seed
- » 1 large container
- » Good-quality potting compost
- » Trowel

1. Water the tomatoes and herbs well before planting.

2. Fill your chosen container with potting compost. Using the trowel, dig a hole in the centre of the compost. Remove the basil from its pot and place it in the hole.

3. Dig two holes either side of the container. Remove the tumbling cherry tomatoes from their pots and place them into the holes.

4. At the front of the container, dig another hole. Remove the oregano from its pot and place it in the hole.

5. At the rear of the container, thinly sow the rocket seed on the surface of the soil. Lightly cover with a little more of the potting compost and water using a spray bottle.

6. Place in a warm and sunny location. Spray mist the seedlings and water the tomatoes regularly. Avoid overwatering the herbs.

7. Thin the rocket (seedlings to approximately 2.5cm (1in) apart.

8. Feed with an organic liquid fertilizer every two weeks throughout the growing season (see pages 38–43).

Cut Flower
Garden

Cut flower garden

I used to love buying flowers from the supermarket or market, but didn't consider how environmentally friendly they were. Most flowers are grown abroad in vast, heated greenhouses (glasshouses) and intensively sprayed with toxic pesticides. After harvesting, the flowers are stored in chilled warehouses then transported in refrigerated trucks and flown thousands of miles to reach their destination. Bunches are wrapped in non-recyclable plastic and once the flowers fade, they are usually sent to landfill rather than composted. That's a huge carbon footprint for a product with a very short lifespan. Lessen the impact by supporting local flower farms or growing your own at home, with the added benefit of providing a vital food source for pollinators.

Whether you buy or grow your own flowers, you can help keep them healthier and fresher for longer by feeding them. The most effective way to do this is to add one tablespoon of sugar and two tablespoons of vinegar (white or apple cider) to the water and mix well. Flowers thrive on sugar, but it can cause bacteria, leading to wilting and slimy leaves. The antibacterial properties of vinegar help to counteract this. Every couple of days, change the water and add the sugar and vinegar before replacing the flowers.

In springtime, supermarkets and florists are full of bunches of hyacinths, tulips, narcissi and daffodils. They look beautiful together in a vase but, when cut, daffodils release a sap that can contaminate the water and infect the other flowers. To help prevent this, always trim your daffodils last and pop them in a container of cold water on their own overnight. In the morning, add them to the rest of your display.

Hints & tips for choosing & growing flowers

THE RIGHT CONDITIONS

Cut flowers thrive in full sun. Avoid windy sites, particularly when growing varieties with tall stems as they may blow over. Prepare any flower beds, raised beds or containers before planting. Flowers need a fertile and weed-free soil, so add plenty of organic matter (see pages 34–37) which helps to improve water retention and drainage.

PLANNING

It's important to consider when flowers bloom as you plan your cutting garden. Ideally, you want to choose varieties that will flower in spring, summer and autumn (fall), providing you with colour and the opportunity to cut throughout the growing season. If you have limited space, choose just one or two varieties that flower each season and think about how you can maximize yield by selecting flowers that bloom continuously.

PLANT TYPES

There are four main types of flower for cutting:

» *Annuals:* Sow the seeds in early spring to harvest in the summer. These varieties die off in the autumn and are often categorized as half-hardy or hardy annuals. Half-hardy varieties are sown indoors and planted out after the risk of frost has passed. Hardy varieties are sown outdoors in the site where they will flower.

» *Bulbs, corms and tubers:* Usually planted in autumn or spring, these sit under the soil, develop roots and then push through the soil to give you a flower. Typically they bloom from spring through autumn, depending on the variety.

» *Biennials:* These live for two years, producing foliage in the first year and blooms in the second. Many self-seed and grow again the following year, repeating the cycle. Sow seeds in early summer and plant in the ground in late summer to flower the following spring.

» *Perennials:* These flower in the summer and autumn and then die back over the winter. However, they do reappear each spring, becoming larger and producing more flowers.

HEIGHT

Be sure to check the height of flowers at maturity when planning your cutting garden. Shorter varieties need to be planted in front of any taller flowers to ensure they get adequate sunshine and water.

STAKING

If you are growing tall flowers, it's a good idea to add supports to the patch, raised bed or container before you plant. You can use bamboo canes and twine to do this. Place a bamboo cane next to each seedling and, as the plant grows, loosely tie the stems to the cane with the twine at intervals of approximately 20cm (8in). For sweet peas, think about making a trellis or wigwam from bamboo canes or, alternatively, invest in a metal obelisk which will last for years and give plants more support in windy conditions (you can often find these at vintage fairs or reclamation yards).

SOWING SEEDS

» *Indoor sowing:* Read the instructions of the seed packet thoroughly before you sow. Choose the right sized container for the variety (see page 63 for how to make one from plastic containers) and fill with good-quality seed compost (see page 61). To germinate effectively, seeds need a good source of light and heat. If you plan to grow lots of flowers, it's a good idea to invest in a grow light, but if you are only growing a couple of trays then a warm, sunny windowsill can work well, too. Before transferring your seedlings outdoors, make sure you let them 'harden off'. This allows them to adjust to the sudden

continued »

change of temperature outdoors. Place your seed trays in a sheltered spot outside and leave them there for at least four hours before returning them indoors. Repeat this daily, increasing the number of hours for at least one week, after which the seedlings can be left permanently outdoors until planting.

» *Direct sowing:* Read the instructions on the seed packet thoroughly before you sow. Place annual seeds into the ground approximately 3–4cm (1¼–1½in) apart in rows or in mass plantings. Annuals like space to grow and they don't want to be in competition with other plants for light, food and water. Small seeds should be planted in small pinches, whereas flowers with large seeds such as sunflowers or calendula should be sown one seed at a time, 3–4cm (1¼–1½in) apart.

PREPARING THE SOIL

Prepare your flower beds, flower patch or raised beds (see page 120) and add some good quality multipurpose potting compost (see page 12) to your soil. If you are going to be sowing seeds directly, then ensure that the soil is a fine consistency to help better germination.

GROWING IN CONTAINERS

Consider the size of pot you will need to use before planting. Choose large, heavy pots for any flowers that have thick stems and large heads, as those plants may need staking.

Begin by adding pebbles or broken pots to help improve drainage. Fill the pot with a good-quality potting compost and either plant up your container with seedlings or sow the seeds directly (see left).

MULCH

Protect any new plants with a mulch of grass clippings, shredded leaves or straw. This helps to suppress weeds and retain moisture.

DAILY CARE

» *Weeding:* Remove weeds regularly, especially during the summer months as they can easily take over the flower bed and stifle plant growth.

» *Pinching out:* This encourages plants to produce more stems and become bushier, which leads to more flower heads. Once the first flowers and three sets of leaves have appeared, take a sharp pair of secateurs (pruning shears) and snip the head off above a set of leaves. Use this technique on annual plants that produce multiple stems and not on single stem varieties.

» *Deadheading:* Remove any spent or damaged flowers to help encourage plants to send up new growth.

» *Watering:* Water regularly and avoid overhead watering when flowers begin to bloom as it can damage the heads. (See pages 18–21 for water-saving ideas.)

» *Pests and diseases:* Remove any affected plants immediately to help prevent diseases spreading (see pages 160–171 for how to deal with pests and diseases naturally).

HARVESTING

Have a bucket of clean, cold water ready before you cut. Aim to harvest flowers before the heads have fully opened and cut either in the morning or evening when temperatures are cooler. Remove any leaves from the lower half of the stems and place the stems in the bucket of water. Leave in the bucket for 2–3 hours before arranging.

A–Z of cut flowers

Here is a list of the most popular flowers that are imported and sold in florists' shops, as well as a few of my own personal favourites. I've included general information on growing, but always read the bulb or seed packet thoroughly before planting as some varieties may have additional requirements.

I have also noted which flowers can be dried successfully – just look for the ❀ symbol.

» *Allium* ❀: (bulb) Plant in full sun, at least twice the depth of the size of the bulb. Alliums are perennial and like room to grow. Plant them in a permanent spot, either in a flower bed or a deep container and they'll reward you year after year with blooms. Bulbs need planting in early to mid autumn (fall), whilst pot-

grown alliums can be planted out from early spring onwards. Flowers in spring and early summer.

» *Ammi majus* ❀: (hardy annual) Grow in full sun or partial shade, flowers from early summer to early autumn. Sow seeds in early spring. Thin the seedlings out, leaving approximately 30cm (12in) between plants. The flowers are similar in appearance to cow parsley and are often used as a 'filler' in floristry. The stems grow tall, so staking may be necessary.

» *Anemone:* (tuber) Plant in autumn in a sunny spot. Soak for 3–4 hours prior to planting. Anemones may need extra protection over winter if the temperature drops below -3°C (26°F). Flowers in early spring.

» *Calendula:* (hardy annual) Sow seeds from early to late spring to flower in late summer and early autumn. Pinching (see page 123) and deadheading make plants more productive. Flowers are edible and delicious added to salads. You can also make your own skincare remedies and soaps with calendula as it has natural antibacterial and antifungal properties.

continued »

» *Carnation:* (hardy annual) Sow seeds from early to late spring. Flowers midsummer to early autumn (fall). Plant in full sun and deadhead regularly to prolong flowering.

» *Cosmos:* (annual) Sow seeds indoors in early spring. Flowers prolifically from midsummer through to mid autumn. Grows very tall, so add supports before planting. Cosmos benefits from pinching (see page 123) and deadheading. It is one of the best flowers to grow in a cutting garden as the more you harvest them, the more they grow.

» *Daffodil:* (bulb) Plant in autumn at twice the depth of the bulb's height. After they have flowered in the spring, allow the leaves to turn yellow and start to wilt before pruning to allow them to effectively photosynthesize. Divide the bulbs every three to four years to increase yield.

» *Dahlia:* (tuber) Plant in spring after the last frost. Lay the tubers horizontally in holes 10cm (4in) deep. Dahlias grow tall and heavy, so will need staking. They benefit

from pinching when they are around 20cm (8in) tall. Flowers in midsummer through to late autumn. After flowering, you will need to dig the tubers up, rinse off any soil residue and allow them to air dry. Wait a couple of weeks after the first frost before doing so as this helps the tuber to go dormant and prevents it from rotting. Pack any tubers in a box or pot and cover with compost. Store in a cool, dry area throughout the winter before replanting in spring.

» *Gypsophila* ✿: (hardy annual) Also known as Baby's Breath and used in wedding bouquets. Plant in full sun. Direct sow in mid spring through to early summer. Flowers from early until late summer. Cut back after flowering to encourage a second bloom.

» *Honesty* ✿: (biennial) Direct sow in early summer in full sun or partial shade. Flowers will appear the following year. Honesty is grown for its translucent and papery seed pods which resemble coins, hence the name money plant. These seed pods look spectacular dried and added to winter flower arrangements.

» *Hyacinth:* (bulb) Plant in autumn in full sun and at least 10cm (4in) deep in the soil. Wear gardening gloves whenever you are handling hyacinth bulbs as they can irritate the skin. Flowers throughout spring. In early autumn divide the bulbs to create new plants.

» *Larkspur* ✿: (half-hardy/hardy annual) Plant in full sun. Sow indoors in late winter and early spring in individual biodegradable pots that will be placed in the soil (larkspur do not like being transplanted). Alternatively, direct sow in mid spring. Pop your larkspur seeds in the freezer for a week or so before sowing as the cold temperature speeds up germination. Thin seedlings approximately 20cm (8in) apart. Flowers throughout the summer. Regular deadheading will prolong flowering. Larkspur petals are used to make biodegradable confetti for weddings as the colours don't fade when dried.

» *Lily:* (bulb) Plant bulbs at least 15cm (6in) deep in full sun. Flowers in late summer and early autumn. Deadhead to encourage further flowers. Once the leaves have died back, cut away any old flower stems.

» *Nasturtium:* (hardy annual) Sow outdoors from late spring to midsummer. They will flower from the summer through to the autumn. Nasturtiums like to climb, so you will need to provide some support for them. They don't like to be overwatered as this can lead to more leaves rather than flowers. Nasturtium flowers and leaves are edible and can be added to salads, soups and even jams (preserves), whilst the pods can be pickled.

» *Nigella* ✿: (hardy annual) Direct sow outdoors in mid to late spring and in a sunny spot. Flowers from midsummer through to early autumn. Regularly deadhead to increase the plant's productivity. Harvest the seed heads after the plant has stopped flowering. The seed pods look beautiful when dried or you can gather the seeds from the pods to use the following year.

continued »

» *Opium poppy* ✿: (hardy annual) Grown for their seed pods, which look spectacular in dried arrangements. Direct sow in the spring once the threat of frost has passed. Flowers in early summer. If you leave the seed heads in place, they will self-seed and produce more plants the following year. I like to cut the seed heads and dry them out, reserving some of the seed to plant in other areas of my garden as well as adding the dried heads to evergreen wreaths and bouquets in winter.

» *Ranunculus:* (bulb) In warmer climates, plant in full sun in the autumn (fall). However, in colder areas, it's best to wait until early spring and after the threat of frost has passed. Before planting, soak the corms for 3-4 hours in room-temperature water. Place them in the soil approximately 5cm (2in) deep and 10cm (4in) apart, with the claw-like end pointing downwards. Flowers appear approximately 90 days after planting.

» *Scabiosa stellata* ✿: (hardy annual) Although these do flower beautifully and bees love them, it's the seed heads that make them special. Direct sow in late spring and early summer, flowers will arrive in late summer through to early autumn. Deadhead regularly to encourage more blooms. Harvest when the seed heads are pale green, and the star-shaped centres are turning black.

» *Sunflower* ✿: (half-hardy/hardy annual) Sow indoors or under cover in early to mid spring. Sunflowers are perfect for kids to grow as they are so easy. You can also repurpose yogurt pots to grow them in – simply add a couple of holes to the bottom of the pot with a pair of scissors for drainage purposes, fill with seed compost and then plant one seed per pot. Water well. Once you can see the roots poking through the holes in the bottom of the pot, they are ready to transplant outdoors. You can also sow seeds directly into containers or flower beds in late spring through to early summer. Sunflowers have tall stems and heavy heads so will require some staking. Flowers from midsummer to early autumn. The heads can be dried and used for feeding birds in the winter.

» *Sweet pea:* (half-hardy annual) Sow seeds in early spring indoors. Soak the seeds in water for 8 hours prior to sowing as this helps to speed up germination. Plant in full sun and in a sheltered spot. Deadhead regularly to prolong flowering.

» *Tulip:* (bulb) Plant in autumn in a sunny spot. Place in a flower bed or container two to three times the size of the bulbs and two bulb widths apart. Cover with soil. Harvest when the flowers are still in bud. Like daffodils, allow the leaves to turn yellow before removing them.

Indoor
Gardening

Indoor gardening

Indoor gardening is fun and accessible to all. Choose one or two plants that have a high success rate such as basil, rocket (arugula) or microgreens (see page 147). Use a good quality potting mix but don't use soil from outdoors as it may contain bacteria or pests.

All plants need a good source of light, but avoid placing them in direct sunlight as they can scorch. Heating and air conditioning can remove all moisture from the air, so spray plants with water at least once a day. Air circulation helps to make plants more resistant to disease, so open windows for a few hours each day if possible, or place an oscillating fan set on low to mimic a gentle breeze. Knowing when to water is vital. When the top of the soil begins to look faded and the container feels lighter, fill a tray with water, then place your container in the water for about 30 minutes until the soil turns

darker. For delicate seedlings, mist using a spray bottle. Put a little fertilizer in the watering tray once a week.

You may have to pollinate certain varieties by hand using an artist's paintbrush or cotton bud (cotton swab). For a self-pollinating plant, gently brush inside each flower and transfer the pollen into the pistil (the middle part of the flower). For other varieties, brush the pollen from the male flower and transfer to the pistil of a female flower (there are handy online tutorials).

Disease prevention is key. Sprinkle ground cinnamon on the top of the compost as it will repel bugs. After you have watered pots, place a saucer of apple cider vinegar nearby as this will attract pests. For any pest infestation, rinse the plant under cold water and shake off bugs. Spray affected parts with natural insecticide.

Edible indoor crops

FRUIT

» *Blueberries:* Self-pollinating, dwarf varieties are best for growing indoors on a sunny windowsill. However, dwarf blueberry bushes can reach heights of 50–100cm (20–40in) so be sure you have the space and the right container to get the best from your plant. Blueberries like a well-drained, acidic soil. Choose a good-quality organic ericaceous potting compost (see page 110) for planting in containers. In order to produce fruit, you'll need to provide your plant with at least 6–8 hours of daylight. Placing a grow light over the plant during the winter months may be necessary to keep it happy. Water only when the top 2.5cm (1in) of soil is dry. Place a large saucer or tray under the pot to collect any drained water. Hand pollination will help the plant to bear fruit (see page 132).

» *Citrus:* Although you can grow citrus from seed, it takes years for the plant to fruit. A better option is to choose a dwarf citrus tree as it provides you with fruit more quickly. Plant in a terracotta pot and use a good-quality citrus potting compost and fertilizer to provide your tree with the best start. Use room-temperature water, and only water when the soil is very dry. Citrus can struggle from a lack of humidity. Daily misting helps but it's also a good idea to place a tray with pebbles under the potted tree as any water that runs through the pot into the pebbles will slowly evaporate and increase humidity. Citrus trees need at least 6–8 hours of sunlight every day and don't like draughts or being placed near a radiator. In warm weather, open the windows to allow air to move around the tree. Flowers will also need to be pollinated by hand.

» *Cucamelons:* These are tiny, delicious balls that taste of cucumber and melon with a hint of lime. They are lovely added to drinks or salads, and they can also be pickled just like cucumbers. They are easy to grow indoors in a pot on a sunny windowsill, but they do grow tall and will need a support system for the vines to cling to. Start by placing the seeds on a damp piece of paper towel. Keep them somewhere warm and the towel damp, but not soaking wet. When you see shoots appearing, transfer your seeds to a container filled with indoor potting compost. Place the container in a sunny spot and water every 4–5 days. Hand pollination is required to help the plant bear fruit. The fruit is ready to pick when it is firm and around the size of a large grape.

» *Strawberries:* You can grow strawberries from their seeds, but it takes many years to get them to produce fruit, so a better option is to buy bare root strawberry plants (see pages 105–106). Look for strawberry varieties that are compact and produce few or no runners. Strawberries have a shallow root system and can grow in small containers and they are

perfect for window boxes and hanging baskets. If you have a large container you can grow several plants together. Before planting, remove any dead leaves or runners. Bind the roots gently together and place the strawberry in the pot with the crown even with the soil. Water well and place in a sunny spot. Check the soil every couple of days and water when the top 2.5cm (1in) of soil is dry. Once the plants begin to produce flowers they will need to be pollinated by hand. Alpine strawberries are also a good option for indoor gardens as they grow upright, and the berries don't require mulching. However, the berries are smaller, and you won't get a huge harvest.

» *Tomatoes:* Small-sized cherry tomatoes are best for growing indoors. Choose bush varieties (see page 100) as these will not need staking or side shooting and can happily grow in a hanging basket or in a large pot. Old yogurt pots are perfect for starting off your tomato plants from seed; just be sure to add a couple of holes in the bottom of the pot for drainage.

continued »

Fill the pot with a good-quality seed compost, then place two seeds in the centre of the pot and on top of the soil. Cover with a little compost and press down gently. Use a watering can with rose attached to water your seedlings and repeat whenever the soil appears dry. Place the seedlings in a warm, sunny spot. Once the seedlings have reached a height of approximately 10–12cm (4–5in), it's time to transplant them. Fill a large container or hanging basket with indoor potting compost. Plant your seedling deep in the pot, just under the first set of leaves. Tomato plants have little hairs all the way up the stem and whenever any of those hairs touch the soil, the plant will put out roots. More roots make for stronger plants. Place in a sunny spot and water well. Check the soil every day and water when the top 2.5cm (1in) of soil is dry.

HERBS

Most herbs will grow well indoors, but some will prefer to be grown from seed whilst others are better transplanted. All herbs require fertilizing every couple of months using a seaweed-based spray (see page 41) and they appreciate good drainage. Add a few stones or pebbles to the bottom of containers to help with this. Before sowing seeds or repotting any plants, pre-moisten the soil first. Place your pot or container in a sunny area. Mist herbs every few days as they thrive in more humid conditions.

» *Basil:* Choose a pot of living herbs from the grocery store or a plant nursery. Repot into a larger container and water well. Alternatively, follow the method on page 145 for how to re-grow basil from a stem cutting.

» *Chives:* Grow from seed, watering only when the soil is dry (but if the tips of the plant start turning yellow, increase the watering).

» *Coriander (cilantro):* Grow from seed and plant in a container at least 20cm (8in) deep. Avoid placing in bright light and water when the soil is very dry.

» *Mint:* Buy a small plant from a nursery and repot into a wide container. Prune regularly to encourage healthy growth and a bushier plant. Keep the soil moist but not saturated.

» *Oregano:* Choose a pot of living herbs from the grocery store or a plant nursery. Repot into a larger container. Let the soil dry slightly before watering again. Pinch off the leaves regularly to increase bushier growth.

» *Parsley:* Grow from seed and place on a sunny windowsill. Water regularly to thoroughly saturate the soil.

» *Rosemary:* Choose a compact bush variety from a plant nursery. Repot into a larger container. Water sparingly as rosemary prefers a drier soil.

» *Sage:* Buy a small pot from a nursery and repot into a larger container. Sage is drought tolerant so water only when the soil is very dry. Sage can suffer with mildew problems and a good way to prevent this from happening is to add a few tiny pebbles to the surface of the soil.

» *Thyme:* Buy a small pot from a nursery or use a grocery-store living herb. Repot into a larger container. Thyme is drought tolerant and should only be watered when the soil is very dry. Prune the stems regularly to promote healthy growth.

VEGETABLES

» *Beans:* For indoor gardens, choose bush varieties as they are ideal for container growing and don't require staking (see page 95 for different types of beans). Beans don't have a large root system, so choose a pot size around 15cm (6in) deep and fill with indoor potting mix. Plant the beans approximately 2.5cm (1in) deep and 5cm (2in) apart, cover with soil and water well. Place in a sunny spot and keep the soil moist to help germination. Once the flowers and beans have appeared, reduce watering until the top 2.5cm (1in) of soil is dry.

» *Carrots:* Choose compact varieties as they need less room to grow and they will mature quickly. Fill a large pot or window box 30cm (12in) deep with indoor potting compost, then moisten and mix with a trowel. Sprinkle a few seeds over surface and lightly cover. Mist well. Place in a sunny spot

continued »

that will receive at least 6 hours of daylight. Use a spray bottle to mist the soil every day. Once seeds begin to germinate you will have to thin them out – using a pair of scissors, gently cut the bottom of the seedlings you want to remove, leaving at least 1cm (½in) between the remaining plants. Repeat this process every couple of weeks, until young plants are 5–7.5cm apart (2–3in). Carrots will be ready to harvest when you see the top of the orange root start to push up out of the soil.

» *Chilli (chile) peppers:* Buy a pepper plant and repot it in a container at least 25cm (10in) high, using indoor potting compost. Allow the pot to dry out completely before watering. Hand pollination is vital if you want your plants to produce more fruit, so use a cotton bud (cotton swab) or fine artist's paintbrush to transfer pollen between flowers (see page 132).

» *Kale:* You can grow kale from seed or buy seedlings. Soak the seeds in a cup of water for about eight hours as this helps to speed up germination. Fill a large

container with indoor potting compost and moisten with water. Sow 3–4 seeds 1–2cm (½–¾in) deep and 15cm (6in) apart. Lightly cover with soil. If you don't want to grow from seed, plant two seedlings in a large container. Place in a sunny spot or use a grow light setup. Water when the soil dries out and if growth is slow, add a little natural fertilizer (see pages 38–43). Harvest the outer and lower leaves first.

» *Mushrooms:* Growing mushrooms can be tricky, but not impossible. There are several different varieties that will grow well indoors but for beginners I would recommend investing in a mushroom kit you can find online or in a garden centre. You can find kits for growing chestnut, shiitake, oyster and portobello mushrooms. They are easy to use as you simply open the kit and spray the mushrooms with water. After a couple of weeks, you should have your first harvest. If you want to try growing your own mushrooms from scratch, then I recommend looking online for specialist growers. Many offer custom-made kits and e-courses too.

» *Radishes:* Plant in a round container or rectangular window box. Fill with indoor potting compost and sow seeds approximately 1–2cm (½–¾in) deep and 5cm (2in) apart, pushing the seeds in gently with your finger. Lightly cover with potting compost and water well. Place in a sunny spot and water regularly. Radishes will be ready to harvest when you see the top of the plant start to push up out of the soil.

» *Salad leaves:* Choose packets of mixed seeds that combine varieties and flavours. Salad leaves have shallow root systems and work well in medium-sized containers and hanging baskets. Fill your chosen container with indoor potting compost, sprinkle the seeds over the top of the soil and lightly cover. Moisten the soil with a spray mister. Salad leaves benefit from regular misting so aim to do this every other day. Place the container in a sunny spot. Harvest when the leaves are around 12cm (5in) high,

picking regularly (known as 'cut and come again').

» *Spinach:* Choose baby leaf varieties for indoor growing. As spinach has a shallow root system, plant seeds in containers approximately 12cm (5in) deep. Fill your container with indoor potting compost and moisten with water. Place your seeds about 1–2cm (½–¾in) deep and at least 12cm (5in) apart. Cover with the potting compost and mist the soil until it's moist. Place in a sunny spot and water when the soil is dry. Harvest when the leaves are 12cm (5in) high.

Sprouting seeds

Like microgreens (see page 147), sprouting seeds are one of the easiest things to grow indoors. They are quick to germinate, and the sprouts are ready to harvest in only a few days. Sprouts are packed full of essential vitamins and minerals and are a good source of both protein and fibre. Unlike microgreens, which are grown in soil, sprouting seeds are grown in water.

You can sprout almost any nut, seed or legume. I like to sprout chickpeas, kale and lentils, as well as the occasional mixed pack of seeds. The table on page 143 lists some of the more common varieties for sprouting along with their soaking and harvesting times.

If you want to grow a lot of sprouts, there are products you can buy. From specialist germinators to clay and terracotta tower sprouters, these will all help you grow and produce more efficiently. You can also find reusable organic hemp bags for growing sprouts, which you dip in water and hang to drain. However, it's simple to sprout seeds using what you already have at home.

EQUIPMENT

» A large clear glass jar (Mason jar or reuse a glass jar from the kitchen such as an instant coffee, pasta sauce or nut butter jar)
» A piece of cheesecloth (muslin), organic and unbleached (cut to cover and overhang the size of your jar) or a clean, dry tea towel (dish towel)
» Rubber band
» Spoon
» A bowl to help the jar stand upside down at an angle

SEEDS

Only use seeds that have been specifically produced for home sprouting as they will have a high germination rate and are subject to strict controls. Search online for organic sprouting seed companies.

HOW TO GROW

The most important thing to consider with sprouting is keeping things very clean as dirty jars and water can cause bacteria to grow. It's simple to prevent this – wash your jars by hand in warm soapy water and rinse well before every use. It is also advisable to wash your hands before and after handling seeds for sprouting.

Place one or two tablespoons of your chosen seed in a clean glass jar. Cover with a couple of inches of cool but not cold water. Stir the seeds with a spoon to make sure they all get wet and push down any that float to the top of the water. Cover the top of the jar with a piece of cheesecloth (muslin) and secure with the rubber band. Soak overnight or for the minimum number of hours listed in the table (see page 143).

Using the cheesecloth as a filter, drain the water. Remove the cheesecloth and fill with a few more inches of fresh cool water. Replace the cheesecloth and swirl the seeds around a little. Prop the jar upside down at a slight angle in the bowl and leave to drain. Repeat this method twice a day.

Whilst growing, leave the sprouts out of direct sunshine. Keep jars on the countertop rather than in a kitchen cabinet as they need air to circulate. Once they have sprouted leaves, it's okay to move them to a sunny windowsill and allow them to 'green' up.

continued »

MOULD PROBLEMS

Should mould appear on any of the sprouts in the jar you will have to discard them and start the process again in a clean jar. Mould can occur if the sprouts are not rinsed thoroughly, so be sure to rinse and drain properly at least twice a day.

STORING YOUR HARVEST

Let your sprouts drain for at least eight hours after their final rinse. Remove from the jar and spread out on a clean, dry tea towel (dish towel). Leave them to air dry for about 30 minutes. Place in a clean glass storage container with lid and pop in the refrigerator. Sprouts should keep fresh for one week.

WHAT TO USE THEM IN

Add sprouts to salads, sandwiches, smoothies, soups or even throw a handful into a homemade pesto or pasta sauce.

Soaking & sprouting times

SPROUT	SOAKING	HARVEST
Alfalfa	8 hours	5–6 days
Beets	8 hours	10–20 days
Broccoli	8 hours	3–6 days
Chickpeas	8–12 hours	2–4 days
Clover	8 hours	5–6 days
Cress	8 hours	4–5 days
Kale	6 hours	3–6 days
Lentils	8 hours	2–3 days
Millet	6 hours	1–2 days
Mung beans	12 hours	3–5 days
Mustard	8 hours	2–3 days
Peas	8 hours	2–3 days
Pumpkin seeds	8 hours	1–3 days
Radish	8 hours	3–6 days
Sesame seeds	8 hours	1–3 days
Sunflower seeds	6 hours	1–2 days
Wheatgrass	8–12 hours	6–10 days

Re-growing from kitchen scraps

Rather than sending kitchen scraps off to landfill or putting them in the compost bin, you can try re-growing them. Certain fruits, herbs and vegetables will successfully root and grow happily in containers on a sunny windowsill.

» *Avocados:* Sadly they are unlikely to produce fruiting plants, but avocado pits do make beautiful houseplants. Start by removing the pit from a ripe avocado and scrape away any flesh that's stuck to it. Wipe dry with a cloth and then gently peel the brown skin away from the pit. You'll notice the pit has several lines running along it and one small hole at the bottom. This hole is where the root will come from. Stick three toothpicks into the top part at a 45-degree angle, evenly spacing them out and avoiding the lines that run around the pit. Use an old jam (jelly) jar or Mason jar and sit the avocado pit

in the jar, top side facing upwards. Fill the jar with clean water. The bottom of the avocado should sit in the water, with the top suspended out of the water with help from the toothpicks. Place the jar in a sunny spot. Top up the jar with water as needed or use fresh water if discolouration occurs. Once you

have lots of roots, transplant the avocado to a container filled with good-quality potting compost. Leave the top part of the pit exposed and water well.

» *Basil:* If you buy a pot of basil from the supermarket, you'll know it often wilts or discolours in a matter of days. This is usually down to the quality of the potting compost the basil has been grown in. To re-grow, simply remove one or two of the healthy stems of basil and place in a glass of water. Change the water every two days and after a couple of weeks you should see new root growth. Fill up a container with good-quality potting compost and plant the basil. Place in a sunny spot and water every two days. You can plant several stems together in one container.

» *Beetroot (beets):* You won't get full grown beetroot by re-growing them, but you can use them for beet greens which are wonderful in salads, soups and smoothies. Save the top part of the beetroot with the roots on and place it cut side down in a shallow dish of water.

Change the water every other day and harvest the beet greens as they grow.

» *Celery and fennel:* Cut off the bottom of the bulb with the root system still attached. Place in a shallow dish of warm water with the base facing down. Pop in a sunny spot, ideally on a windowsill. Change the water every couple of days to prevent the bulb rotting or going mouldy. After a short while you should see signs of new growth. Remove the base from the water and transplant into a container with a good-quality potting compost. Cover the base, leaving the new growth above the soil. Water well.

» *Coriander (cilantro):* Like basil, you need a pot of coriander from the supermarket to do this. Remove a few healthy stems of the coriander from the pot and place in a small jam jar or glass filled with water. Place on a sunny windowsill and change the water every other day. Once the stems have produced a good number of new roots, transplant to a container filled with good-quality potting compost.

continued »

» *Ginger:* Make sure it is firm and smooth skinned, and ideally with some buds already forming. Soak the root in warm water overnight before planting. You will need to have a container at least 40cm (16in) deep to grow ginger. Place the root in your container about 5cm (2in) below the lip of the pot and cover with a further 2cm (¾in) of potting compost. If you have a piece of root ginger with buds, place the buds facing upwards. Water well. Once shoots start to appear, mist regularly. Ginger is ready to harvest after one or two years, but it's worth the wait!

» *Lemongrass:* Place the leftover lemongrass root in a glass filled with about 2.5cm (1in) of water. Place in a sunny spot and top up with water if it starts to evaporate. After a couple of weeks, you should see the lemongrass has grown new roots. Transplant to a container filled with good-quality potting compost.

» *Pak choi/bok choi:* Place the root end in a shallow dish of warm water. Pop in a sunny spot and after a couple of weeks, replant in a container with a good-quality potting compost.

» *Peppers:* Save any seeds from chillies (chiles) or (bell) peppers. Prepare a tray with good-quality seed compost (see page 61 for how to make your own). Scatter the seeds on top of the compost and lightly cover. Fill a shallow tray with water and place the seed tray over the top to allow the water to soak up slowly. After 8–10 days, you should have several seedlings ready to transplant. Gently pull out each seedling and replant in individual containers filled with good-quality potting compost.

» *Spring onions (scallions):* You need to make sure that any spring onions you want to use still have a few roots on them, otherwise they won't re-grow. Cut 2.5cm (1in) above the roots of the spring onion, then place the roots in an old jam (jelly) jar. Cover the roots with fresh water, leaving the stems dry. Change the water every second day. After 8–10 days, the spring onions will start re-growing and will be ready to eat. Cut and use what you need and then repeat the process.

Microgreens

My favourite plants to grow indoors are microgreens. They are simply plants that have not reached maturity and are harvested around 14 days after germination. These tiny greens are perfect for beginners as they are simple to grow, germination rates are high, and they taste delicious.

Microgreens are packed full of vitamins, minerals and antioxidants. They are wonderful added to sandwiches, salads, smoothies and soups, made into pesto or used as a garnish.

EQUIPMENT

» *Growing medium:* Indoor potting compost (see page 61).

» *Containers:* Seed trays are ideal for growing microgreens. Old seed trays need to be washed in warm, soapy water, then rinsed and left to air dry before use. If you plan to buy new seed trays, opt for ones made from bamboo which will last for several years. Once they come to the end of their lifespan, they can be broken up and added to the compost bin. Alternatively, make your own container from plastic produce cartons or foil takeaway (takeout) dishes (see page 63). You can also grow microgreens on a hemp grow mat, which are soilless and made from natural, sustainable fibres. Hemp mats are biodegradable and compostable and have good germination rates. If you are growing microgreens on hemp mats you will require a mister for watering and a seed tray to place the mat in.

» *Seeds:* You don't need to buy special microgreen seeds for this purpose as regular vegetable seeds will do the job. Most online seed growers offer collections of microgreen varieties, including mixed packets of seeds. These can be a good option as they generally mix seeds that have similar germination rates and can be harvested at the same time. You can also find microgreen seeds for sale on specialist microgreen growers' websites, but you may have to pay a little more for them.

continued »

HOW TO GROW

You need two seed trays for each crop, one to grow in and the other to cover during germination. The tray used to cover should not have any holes in it as it needs to block the light. Fill one tray with indoor potting compost and lightly level the surface. If using a hemp mat, place that in the seed tray. Using a watering can with a rose, lightly water the surface of the soil or the hemp mat. Generously scatter your chosen seed across the top of the soil or mat and lightly water again. Stack the second seed tray on top of the first and place somewhere dark. After a few days, remove the top seed tray and move the tray with the microgreens to a brighter location or under a grow light. For soil-grown microgreens, water from below whenever the soil looks dry and for hemp mat-grown crops, mist from above once a day. Most microgreens are ready to harvest when they reach 5–7cm (2–3in), although sunflower and pea shoots should be harvested when 10–12cm (4–5in) high.

MICROGREEN VARIETIES

» *Amaranth:* mild earthy flavour. Germination 4–6 days; harvest 10–20 days after germination.

» *Beetroot (Beets):* sweet and earthy flavour. Germination 2–3 days; harvest 12–8 days after germination.

» *Broccoli:* mild broccoli flavour. Germination 4–6 days; harvest 10 days after germination.

» *Coriander (cilantro):* earthy, citrus flavour. Germination 6–8 days; harvest 12–16 days after germination.

» *Fennel:* aniseed flavour. Germination 4–6 days; harvest 14 days after germination.

» *Kale:* broccoli flavour. Germination 2–3 days; harvest 8–12 days after germination.

» *Kohlrabi:* mild cabbage flavour. Germination 2 days; harvest 8–12 days after germination.

» *Mustard:* spicy, peppery flavour. Germination 4–6 days; harvest 10–20 days after germination.

» *Nasturtium:* spicy, peppery flavour. Soak the seeds for one hour before adding to the growing medium. Germination 5–7 days; harvest 5–10 days after germination.

» *Pak choi:* mild cabbage flavour. Germination 5–6 days; harvest 9–14 days after germination.

» *Parsley:* mild parsley flavour. Germination 7–15 days; harvest 10–15 days after germination.

» *Pea shoots:* fresh pea flavour. Put pea seeds in a bowl and soak in cold water for 4 hours. Drain, then follow the instructions opposite for how to plant. Germination 4–6 days; harvest 14 days after germination.

» *Radish:* fresh, spicy and peppery flavour. Germination 3–5 days; harvest 9 days after germination.

» *Red cabbage:* mild cabbage flavour. Germination 4–6 days; harvest 12–16 days after germination.

» *Rocket (arugula):* spicy and peppery flavour. Germination 4–6 days; harvest 12–16 days after germination.

» *Sorrel:* lemon flavour. Germination 4–6 days; harvest 10–20 days after germination.

» *Sweet basil:* peppery, sweet flavour with a hint of mint. Germination 4–6 days; harvest 10–20 days after germination.

» *Sunflower shoots:* mild flavour, crunchier than pea shoots. Put the seeds in a bowl and soak in cold water for 12 hours. Drain, then follow instructions opposite for how to plant. Germination 4–6 days; harvest 8–10 days after germination.

» *Thai basil:* spicy and peppery flavour. Germination 4–6 days; harvest 10–20 days after germination.

Houseplants

Buying houseplants can turn into an obsession. There are so many beautiful varieties, from succulents trailing over bookshelves and concrete pots filled with spiky cacti to terrariums with tiny air plants (see page 153). It is all too easy to be influenced by images of homes filled with plants in magazines or social media, but it is essential to choose plants that will work for the conditions in your home. If you can't provide them with what they need to thrive, you are setting them up to fail.

There are a few things to consider before buying any houseplants:

» *The light in your home:* Although houseplants tend to prefer bright, indirect light, some can tolerate low light levels whilst others like succulents and cacti do well in warm, bright sunshine. Observe the light in your home. Do you have a sunny windowsill or a dark, shady area? Study how the light changes throughout the day.

» *Think about the space you have:* Large specimens need room to grow and repotting into larger containers. Is there a lot of moisture in the air, or is it very dry? Bathrooms and kitchens have different conditions to living rooms and bedrooms. Do you have time to water regularly or are you away from home a lot?

Once you have answered these questions, pop along to your local plant store or search online for a specialist grower. They can offer expert advice on the right plant for your space as well as tips on how to water, fertilize and propagate correctly.

HOW TO LOOK AFTER YOUR HOUSEPLANTS

» *Potting mix:* Most indoor plants will be happy growing in a good quality, peat-free multipurpose potting compost (see page 12) although others such as orchids, citrus, cacti and succulents will require a specialist compost.

» *Watering:* All indoor and outdoor plants prefer rainwater to drink, so wherever possible, use stored rainwater for watering (see pages 18–21 for water saving ideas). To make tap water more palatable for plants let it sit for a few hours to achieve room temperature and for any chlorine in the water to evaporate. To check if your houseplants need watering, insert a finger into the soil to a depth of around 5cm (2in) and if the soil feels completely dry, it is time to water. Alternatively, insert a bamboo skewer into the soil. If the skewer comes out clean, it means the compost is dry and it is time to water. However, if there are fragments of soil stuck to the skewer, there is enough moisture in the pot and the plant does not need watering.

» *Humidity:* Some houseplants require a damp and humid atmosphere to survive. To increase the level of humidity, mist plants every day with water or place pots on a saucer or tray filled with pebbles. Fill the saucer or tray with water halfway up the pebbles. Empty and refill the saucer or tray with clean water regularly to prevent pest infestations.

» *Fertilizing:* Most houseplants will appreciate a regular feed of an organic liquid fertilizer throughout

continued »

the growing season. For zero waste options, use the leachate from a worm farm (see page 55) or make a batch of banana peel tea (see page 38). Cacti succulents, citrus and orchids require specialist fertilizing products. Be mindful of over-fertilizing as this can burn the roots of plants. Slow release fertilizers are added to potting mixes, providing plants with enough nutrients for six weeks. Always check the manufacturer's label for the correct dosage level and instructions for use. I find a repurposed plastic milk bottle helpful for measuring the ratio of fertilizer to water before I add it to my watering can.

» *Cleaning:* It is important to keep your houseplants clean. Dust stops light getting to the leaves and blocks their pores, reducing a plant's ability to photosynthesise. Wipe down large glossy leaves with a soft, damp cloth. For plants that are spiky or furry, use a cotton bud or small artists paintbrush to gently brush any dirt away. In warm weather, pop your houseplants outdoors to enjoy a natural rain shower. Alternatively,

give them an indoor shower. Use lukewarm water and leave to drain freely.

» *Pests and diseases:* Like all plants, houseplants can come under attack from pests and disease. Please refer to pages 162–165 for types of pest and disease and pages 168–170 for natural ways to deal with pests and diseases.

Be aware that many houseplants can be toxic to humans and pets if ingested. To ensure that children and furry friends stay safe, place houseplants out of harm's way.

To make growing houseplants more sustainable, avoid buying plants that only live for one season such as poinsettias and sprayed cacti. Learn how to propagate and grow new plants as gifts. Reuse takeaway cups and tins to grow baby plants. Search online for videos that teach you how to take stem and leaf cuttings. Look for local plant-swapping events where you can exchange baby plants and cuttings, tools, seeds and pots.

Make an upcycled terrarium

A terrarium is a wonderful way to welcome nature indoors. You can purchase readymade glass terrariums, but it's very easy to upcycle a clear glass jar or a bottle and make your own. Small-sized jar terrariums filled with succulents make wonderful gifts for friends or teachers – just remember to attach a label with care instructions.

There are two ways to make a terrarium: closed container or open container. A closed container has a lid or a stopper, sealing in moisture and raising the humidity level. An open container is exposed to the air and is better suited for growing desert-loving plants.

A glass cookie jar, a large Mason/ preserving jar or a vintage bottle all make ideal vessels for closed container terrariums. You will also need a removable lid or stopper for a closed container.

For open container terrariums, repurpose an old jam (jelly) jar, coffee jar or a glass vase.

Many plant varieties will grow happily in a terrarium. For closed container terrariums, choose humidity-loving plants like bromeliads, ferns, orchids and moss. For open containers, look for plants that enjoy drier conditions, such as succulents, cacti and air plants. Search online for specialist nurseries that grow baby plants for terrariums.

I use a layer of alpine grit for drainage purposes, but pebbles or gravel will also work well. If you are growing cacti or succulents, invest in a specialist potting compost. For other plants, use a good-quality indoor potting mix. Closed containers require a layer of activated charcoal to remove any bad odours and to prevent the growth of fungi.

continued »

YOU WILL NEED

» Glass vessel

» Alpine grit

» Activated charcoal (for closed containers)

» Indoor potting mix or cactus compost

» Large spoon

» Selection of small plants (see right)

» Spray bottle with mister or teaspoon

1. Wash your chosen container in warm, soapy water. Rinse, then leave to air dry.

2. Place a thin layer of alpine grit in the bottom of the container. If you are using a closed container, cover the alpine grit with a thin layer of activated charcoal.

3. Next, add a layer of indoor potting mix or cactus compost approximately three times the depth of the layer of the alpine grit.

4. Using the large spoon, hollow out a hole in the potting mix for each plant. Place plants in the holes and pat the soil firmly around the roots of each plant.

5. Cover the compost with a layer of alpine grit.

6. For closed containers, water the terrarium with the spray misting bottle, then attach the lid or stopper.

7. For open containers, water each plant with a teaspoon of water.

PLANTS FOR TERRARIUMS

Closed Containers:

» Earth Star plant (*Cryptanthus bivittatus*)
» Polka dot plant (*Hypoestes*)
» Nerve plant (*Fittonia*)

Open Containers:

» Echeveria 'Blue Sky'
» Zebra cactus (*Haworthia*)
» Donkey's Tail (*Sedum burrito*)

CARING FOR A TERRARIUM

» *Caring for a closed container terrarium:* Check the soil moisture level every two weeks. If the soil is dry, water well with a spray misting bottle. Avoid placing your terrarium in direct sunlight as it can get too hot and burn the leaves and stems of your plants. Remove the lid or stopper once a month to release any trapped condensation.

» *Caring for an open container terrarium:* Keep out of direct sunlight. Check the soil moisture level every two weeks. If the soil is completely dry, water each plant with a teaspoon of water

Eight houseplants to reduce indoor air pollution

Houseplants are the superheroes in our homes. They have the power to reduce levels of indoor air pollution, eliminating toxins such as benzene, formaldehyde and trichloroethylene from the air. These are commonly found in furniture, paint, cleaning products, plastic and carpets. Long-term exposure can cause respiratory problems, severe headaches and dizziness, so include a few houseplants in your home to provide cleaner, healthier air.

The houseplants listed here are all easy to care for and good for reducing indoor air pollution.

SNAKE PLANT
(*Sansavieria trifasciata/Mother-in-law's tongue*) Slow growing and one of the best plants for improving indoor air pollution. Prefers bright light but can handle some shade too. Snake plant like to have its roots crowded, so choose a tight-fitting container. Use a good quality cactus compost and feed once a month from spring to autumn. Water sparingly in autumn and winter. Propagate from leaf cuttings.

DEVIL'S IVY
(*Epipremnum aureum/Golden Pothos*) Trailing vine suitable for hanging planters or it can be trained to grow up a moss pole. Ideal for rooms with high levels of moisture like kitchens and bathrooms, it needs occasional misting if located elsewhere. Does not require a lot of watering but will appreciate a monthly feed of liquid fertilizer in spring and summer. Propagate in spring from stem cuttings.

CORN PLANT
(*Dracaena fragrans/Dragon plant/Dragon Tree*) Happy in light shade

to bright, indirect light. Does not need a lot of watering but does like to be misted every couple of days. Good for bathrooms. Feed monthly with organic liquid fertilizer in spring and summer. Propagate in spring from stem cuttings.

SPIDER PLANT
(Chlorophytum comosum 'Variegtum')
Tolerant of both light and shade. Avoid direct sunlight which can scorch the leaves. Feed monthly with organic liquid fertilizer in spring and summer, but dilute any liquid fertilizer by half the recommended amount as it prefers a weaker feed. Spider plants develop baby plantlets regularly; simply pinch these off and replant in compost to create new plants.

WEEPING FIG
(Ficus benjamina)
Likes bright, indirect light. Prefers moist soil and absorbs a lot of water through its leaves. Mist frequently. Rotate the pot weekly to ensure the plant grows evenly on all sides. Feed monthly with organic liquid fertilizer in spring and summer. Propagate from leaf cuttings.

RUBBER PLANT
(Ficus elastica/Rubber Tree/Rubber fig)
Likes bright, indirect light. Dust and mist regularly to keep leaves glossy. Feed monthly with organic liquid fertilizer in spring and summer. Propagate from leaf cuttings.

JADE PLANT
(Crassula ovata/Money tree/Friendship tree) Prefers a warm, sunny window-sill. Water sparingly in growing season. Feed once in spring and again in summer. Use good-quality cactus compost. Propagate from leaf cuttings and let cuttings dry out before potting on.

PEACE LILY
(Spathiphyllum wallisii/white sails/spathe flower) Tolerant of most light conditions but will only flower in a brighter location. Thrives in humid conditions and particularly useful in a bathroom to reduce the amount of mould spores in the air. Mist often or place on a tray with a few damp pebbles. Feed monthly with organic liquid fertilizer in spring and summer. Propagate by division in spring.

Pests & Diseases

Pests & diseases

Gardening takes a lot of time and hard work, so it can be demoralising to discover a plant succumbing to a pest or disease. In the organic garden, prevention is key: look after your soil as healthy, fertile growing conditions help plants to thrive and diseases to wane. Give plants room to breathe as good air flow helps to prevent disease. Choose plants that work with your soil type (see page 30) as those that struggle to thrive will be more susceptible to pests and diseases. Keep your plants well-watered and avoid over-fertilizing.

Crop rotation is also vital in preventing pests and diseases from taking hold, so rotate annual vegetable crops each growing season. Choose pest- and disease-resistant varieties for planting.

Be vigilant; check netting is securely fixed and that there are no holes (see page 23). After periods of rainfall, check traps and barriers for slugs and snail invasions. Replace any damaged traps or barriers immediately.

Encourage natural predators into the garden. Hedgehogs love snails and beetles, birds will feast on caterpillars and ladybirds (ladybugs) will enjoy a heathy diet of aphids. Read the chapter on biodiversity to help encourage more wildlife into the garden.

Add companion plants (see page 166) to your vegetable bed. Pests use their sense of smell to find a food source, but several herbs and flowers can mask the scent and deter infestations.

If you have problems with pests or diseases, identify what is causing harm to your plant before applying a treatment. The most common problems and how to deal with them naturally are on page 162.

Common pests & diseases

Here are some of the most common pests and diseases found in the vegetable garden and simple, natural ways to deal with them. If you are struggling to identify what is causing harm to plants or crops in your garden, ask your local nursery for help or search online for help with identification.

APHIDS

These sapsucking insects cause stunted growth and curled or mottled leaves. They secrete a sticky substance known as honeydew, on which a black sooty mould can grow. Aphids are tiny, soft-bodied creatures and can often be found in clusters on buds or the underside of young leaves. They are mostly green or black in colour (known commonly as greenfly or blackfly) but you can also find grey or brown aphids.

» *Preventative measures/treatment:* Blast them off with a jet spray of water from the hosepipe. Spray any infestation with the Castile Soap Insecticide Spray (see page 169). Encourage ladybirds (ladybugs) into the garden as they are the natural predator of aphids (see page 76).

BLIGHT

This is a fungal disease that spreads from windblown spores and affects potatoes and outdoor tomatoes. Mostly prevalent when conditions are warm and wet, plants that have been infected will turn brown and shrivel before collapsing. If the disease reaches potato tubers or tomato fruits, they will rot.

» *Preventative measures/treatment:* Choose blight-resistant potato and tomato varieties. Grow first and second early potatoes (see page 102), which can be harvested before blight arrives. Grow tomatoes in greenhouses (glasshouses) as these offer more protection. At the first sign of infection in potatoes, remove all the plant above soil level to prevent the disease affecting the tubers. Destroy any infected crops, including those in the soil. Do not compost.

CABBAGE ROOT FLY

This resembles a housefly and lays its eggs on the surface of the soil. When the eggs hatch, the larvae tunnel into the roots of brassicas, radish, swede (rutabaga) and turnip. Crops will wilt and shrivel, and some may die.

» *Preventative measures/treatment:*
Use protective netting (see page 23). Make collars for broccoli, Brussels sprouts, cauliflowers and cabbages to deter the fly laying eggs beside the plant roots. Repurpose cardboard boxes or use pre-cut card circles that sit under pizzas from the grocery store; carpet offcuts or doormats also work well. Use a saucer or a CD as a template to cut out circles. Cut a straight line from the circle's lower edge to the centre, then cut one small line diagonally to the left and the same to the right to form a Y shape. Place the collar around the cabbage stem and push down gently. Use a stapler to overlap the cut edge and seal the collar. Water to flatten the card or carpet offcut. The collar should sit on the surface of the soil. Add cardboard collars to the compost bin at the end of the season.

CABBAGE CATERPILLARS

The caterpillars from cabbage white butterflies (large and small) and cabbage moths feast on leaves of broccoli, cabbages, nasturtiums and Brussels sprouts. Large cabbage white caterpillars are yellow, black and hairy, whilst the larvae from small cabbage white are hairy and pale green. Cabbage moth caterpillars have smooth bodies that are greenish brown in colour.

» *Preventative measures/treatment:*
Plant nectar-rich flowers and herbs to attract natural predators that feed on the larvae (see plants for pollination on page 78). Cover plants with butterfly netting from early spring through to autumn (fall). Hand pick and remove any larvae or eggs hiding under leaves.

CARROT FLY

Black, small-bodied fly that lays eggs just under the soil surface of carrots, celeriac, celery, parsley and

continued »

parsnips. The hatched maggots devour small root hairs, then tunnel into the roots of vegetables, leaving them inedible.

» *Preventative measures/treatment:* cover with insect-proof netting, sow sparsely to avoid thinning seedlings later or companion plant with chives (see page 166).

CODLING MOTH

A small, brown moth that lays its eggs on developing fruits. The eggs hatch into larvae that burrow through the fruit and into the core. Codling moths favour apple and pear trees, but are also known to harm quince and walnut trees.

» *Preventative measures/treatment:* Use a pheromone trap. Hung from a fruit tree, the trap lures males using the scent of the female moth. Capturing the males reduces the success rate of mating with a female, resulting in less egg laying. Any fruit that gets damaged by the codling moth should be removed and added to the compost bin.

POWDERY MILDEW

Powdery spots of white or grey fungus that appear on the leaves and stems of plants, powdery mildew is a disease that affects apples, courgettes (zucchini), cucumbers, grapes, peas and strawberries. As the fungi block spores and light, photosynthesis will be impaired, impairing the quality of fruits and vegetables.

» *Preventative measures/treatment:* Remove and dispose of infected leaves (do not add to the compost bin). Spray the plant with one of the antifungal sprays on pages 169–170. Spray in the early morning and do not fertilize until the infection is over. Wash any garden tools in warm, soapy water after using.

SLUGS & SNAILS

Every gardener's nemesis!

» *Preventative measures/treatment:* Circle tender plants with crushed eggshells, coffee grounds or pine needles as slugs and snails find it hard to crawl over coarse materials. Encourage wildlife – birds, frogs, hedgehogs etc are natural predators for slugs and snails. Place 'trap' plants like chervil and marigolds next to vulnerable slug-loving plants (see page 167). Hand pick and remove culprits in the evening or early morning. Add seaweed mulch (see page 36) as the salt scent will deter them and, once dry, seaweed is too hard for slugs and snails to traverse. Wool pellets also work; scatter around fruits and vegetables and water well. The pellets swell up, revealing tiny fibres that are an irritant to slugs. Scoop out half a grapefruit and place cut side down near tender plants overnight. In the morning, remove slugs inside the grapefruit.

VINE WEEVIL

Adult beetles lay their eggs near the base of plants, and these hatch into creamy-white grubs with brown heads that live in the soil. These grubs feast on the roots, causing plants to wilt and die. Vine weevil thrive in container-grown plants and are fond of tender perennials and strawberries. The adults eat the edges of a leaf and this can be a sign that you have a vine weevil infestation.

» *Preventative measures/treatment:* Add a layer of horticultural grit or pea shingle to the surface of the soil in containers to deter the adult weevils from laying eggs.

Hand pick adult vine weevil from leaves in the evening. Encourage natural predators, such as birds, frogs and hedgehogs (see page 86). If you suspect a plant has been eaten by vine weevil, remove it from its container and look at the root system. Grubs are often still present so remove these and add to bird tables or garden ponds. Vine weevil killer treatments made with natural nematodes (see page 171) are also a good option for the organic gardener.

WHITEFLY

Also known as greenhouse whitefly, these tiny sap-sucking insects lay eggs on the underside of leaves, depositing white wax. Adults excrete honeydew on which black mould can grow. Attacks plants indoors and outdoors; favourites are aubergines (eggplant), cucumbers, (bell) peppers, pumpkins and strawberries.

» *Preventative measures/treatment:* Blast them off with a jet from the hosepipe. Spray any infestation with the Castile Soap Insecticide Spray on page 169. Encourage ladybirds (ladybugs) into the garden.

Companion planting

Companion planting is used to improve crop production, maximize space, increase pollination, reduce weeds and save water. However, the greatest benefit to be gained from companion planting is pest control.

Here are eight of the most useful companion plants in the vegetable garden. Herbs mask the scent of fruits and vegetables from pests, whilst improving the production and flavour of the crops they are protecting. Flowers in the veg bed entice pests away from crops while attracting pollinating insects. Include a few companion plants in your vegetable bed at the start of each growing season.

COMPANION PLANTS

» *Basil:* Plant next to tomatoes in a greenhouse (glasshouse) and to attract whitefly away from crops. Aubergines (eggplant), lettuce and (bell) peppers also benefit from basil being planted close by as it helps to increase plant health and productivity.

» *Borage:* Plant near strawberries as it is said to improve their flavour. It is a wonderful plant for encouraging pollinators into the garden. Good next to summer and winter squash, too.

» *Chives:* Plant alongside carrots as the smell from the flowers will repel the carrot fly. They will also improve the flavour and growth of carrots. Chives deter aphids and are useful planted next to apples, berries, peas and tomatoes.

» *Marigold (tagetes):* Plant next to asparagus, melons, potatoes, squash and sweetcorn to repel whitefly and aphids. Marigolds feed pollinating insects and increase vegetable production.

» *Lavender:* With its heady fragrance lavender can repel many pests. Plant next to celery and members of the brassica family.

» *Nasturtiums:* Deters aphids when planted next to members of the bean family. Nasturtiums are fantastic planted alongside cucumbers as they repel cucumber beetles and prevent bacterial wilt. They should also be planted next to apples, potatoes, pumpkins, radishes and squash.

» *Rosemary:* Disguises the scent of cabbages and other brassicas from moths and cabbage white butterflies. Deters pea and bean weevil, and bean beetles, with its strong scent. A handy tip is to scatter rosemary cuttings around the tops of growing carrots to mask their scent from carrot flies.

» *Sage:* Plant next to the brassica family as it deters cabbage whites and moths. Tomatoes excel when planted next to sage, but avoid placing this herb near cucumbers.

TRAP CROPS

Trap crops (also known as decoy or sacrificial crops) are a good pest deterrent in the vegetable garden. You can either use the same crop (by planting the trap crop first, then following with the harvest crop) or use a type of crop that is more attractive to pests. When the trap crop becomes infested, it can be removed and destroyed. Good trap crops are nasturtiums for aphids, chervil and marigolds for slugs, and mustard for cabbage caterpillars.

Natural pesticides & fungicides

Pesticides used in gardens and food production cause harm to wildlife and the environment. Plants treated with pesticides are ingested by insects, which in turn get eaten by birds, hedgehogs and small mammals. Pollinating insects like honeybees are affected too, leading to a decline in crop pollination and reproduction. In some cases, entire wildlife habitats are under threat as there may no longer be a viable food source. There are many types of pesticide available, from weedkillers and bug sprays to lawn treatments, each used to achieve a 'healthy' garden. Yet these treatments are anything but healthy. Pesticides do not discriminate; they kill all insects, including those required for pollination. They also destroy vital nutrients in the soil.

It's simple to make pesticides that are environmentally friendly and safe for wildlife. These non-toxic sprays work on most pests and diseases. Before applying, water your plants well. Do a test spray on a small area at least 48 hours before full application. Apply in the morning and not in full sun. If leaves turn yellow or appear burnt, wash with cold water.

For information on ingredients used, see pages 14–15. Water from the cold tap (faucet) is fine, but if you live in an area of hard water, use distilled or bottled water.

Castile soap insecticide spray

Use on aphids, earwigs, mealybugs and whitefly

YOU WILL NEED

» 500-ml (16-oz) bottle with spray attachment
» Water
» Teaspoon
» Liquid castile soap
» Ground cinnamon (optional)

1. Remove the nozzle from the spray gun and fill the bottle with water.

2. Add 1 teaspoon liquid castile soap and ¼ teaspoon ground cinnamon (optional).

3. Attach the spray nozzle and shake to combine.

HOW TO APPLY
Spray the tops and backs of the leaves until well covered. Repeat every three days until infestation ceases.

Neem oil

Use as an insecticide and fungicide.

YOU WILL NEED

» 500-ml (16-oz) size bottle with spray attachment
» Water
» Teaspoon
» Organic cold pressed neem oil
» Peppermint essential oil
» Liquid castile soap

1. Remove the nozzle from the spray gun and fill the bottle with water.

2. Add 1 teaspoon neem oil, 8 drops peppermint essential oil and 1 teaspoon liquid castile soap.

3. Attach the spray nozzle and shake to combine.

HOW TO APPLY
Spray the tops and backs of the leaves until well covered. Repeat weekly until infestation ceases.

Bicarbonate of soda (baking soda)

Use on aphids, earwigs, mealybugs and whitefly

YOU WILL NEED

» 500-ml (16-oz) size bottle with spray attachment
» Water
» Teaspoon
» Bicarbonate of soda (baking soda)
» Liquid castile soap

1. Remove the nozzle from the spray gun and fill the bottle with water.

2. Add ½ teaspoon bicarbonate of soda (baking soda) and 1 teaspoon liquid castile soap.

3. Attach the spray nozzle and shake to combine.

HOW TO APPLY

Spray the tops and backs of the leaves until well covered. Use weekly until infestation ceases.

Chives & flowers

Chives have antibacterial and antifungal properties. Use on powdery mildew and black spot

YOU WILL NEED

» Large handful of chive leaves and flowers
» Boiling water
» Bowl, funnel and strainer
» 500-ml (16-oz) size bottle with spray attachment
» Liquid castile soap

1. Place chive leaves and flowers in a bowl and cover with 500ml (16fl oz/2 cups) boiling water. Leave to cool.

2. Attach funnel to the spray bottle, then place the strainer over the top of the funnel. Strain the mixture through.

3. Add 1 teaspoon liquid castile soap to the bottle. Attach the nozzle and shake to combine.

HOW TO APPLY

See left.

Other ways to deal with pests

The scent of peppermint repels many aphids, slugs and snails. Hang fabric strips soaked in peppermint essential oil near crops (one drop of undiluted organic essential oil per strip). Always check with your veterinarian before using any essential oil around pets.

Birds should be encouraged into the garden as natural predators but there are a few that cause more harm than good. Pigeons are often the worst at demolishing seedlings. Deter them by hanging wind chimes above the veg garden. Cover crops with netting or place a plastic bottle cloche (see page 23) over seedlings. String old CDs or DVDs from the branches of fruit trees as birds are not keen on sudden movements or flashing lights.

Covering crops with netting is one of the most effective ways to deter pests. Unfortunately garden netting is made from non-renewable plastics so buy a high-quality, rot-proof UV stabilized plastic. This lasts for a long time and will not need to be sent to landfill at the end of the season. Choose butterfly netting that allows access for beneficial pollinators but prevents larger pests from getting in.

In the greenhouse pests can be stopped using biological controls. There are two methods: adding beneficial nematodes via water to the soil or releasing natural predators into the environment. Nematodes kill pests by injecting them with deadly bacteria. You can find nematodes to attack slugs and vine weevil, among others. Predators target specific pests but will not harm beneficial insects. Biological controls do not damage plants and pests do not develop resistance to them. The type of pest determines whether you need a nematode or predator. Look online for biological control suppliers.

Green sources of inspiration

INSTAGRAM

@small_sustainable_steps Amanda has transformed her urban garden into a highly productive edible garden that feeds and nourishes her family and the wider community. Helpful tips and practical solutions for accessible gardening.

@annagreenland An organic gardener for Michelin-starred chefs, Anna shares her passion for growing edibles on Instagram. Her how-to videos are particularly useful, and she includes recipe ideas for how to cook with what you have harvested.

YOUTUBE

Lovely Greens Tanya's wonderful channel is packed full of handy tips for the organic gardener. She also shares recipe ideas and how to make your own soap and natural beauty products from flowers and herbs grown at home.
YOUTUBE.COM/USER/LOVELYGREENSTV

CaliKim Simple videos for growing organic fruit and veg in small spaces. CaliKim also does regular live chat videos and answers all those common gardening queries.
YOUTUBE.COM/USER/CALIKIM29

PODCASTS

Nature & Nourish with Becky Cole Lovely podcast embracing the seasons. Hosted by Becky, an ethical farmer in Northern Ireland, it includes foraging tips, what to grow and when, natural beauty makes and herbal medicine.
BECKYOCOLE.COM

The Organic Gardening Podcast Monthly podcast that breaks down what to do each month and answers listeners' questions. Interviews with well-known organic gardeners and discusses topics such as vegan gardening, attracting wildlife and seed saving.
GARDENORGANIC.ORG.UK/PODCAST

ONLINE SOURCES

Franklyn + Vincent Sarah's blog has long been a favourite of mine. She shares tips and DIY ideas, including a wonderful modern planter on wheels. FRANKLYNANDVINCENT.COM

Vertical Veg A fantastic resource for those growing edibles in containers and on rooftops and balconies. There are video tutorials and inspiring stories from other container gardeners around the world. VERTICALVEG.ORG.UK

Gardenista Garden design tips, how-to-grow advice and beautiful garden tours from around the world. GARDENISTA.COM

APPS

My Soil View a map of the soil in your local area and find information on soil depth, texture, pH, temperature and organic matter contained in the soil. UK and Europe. BGS.AC.UK/MYSOIL/

RHS Grow Your Own Detailed growing, sowing and harvesting instructions for over 100 edibles. It has alerts to remind you what to do and when, as well as advice for tackling pests and diseases. RHS.ORG.UK/ADVICE/GROW-YOUR-OWN/APP

OTHER RESOURCES

Share Waste An Australian initiative that connects people to share their kitchen waste with those who have the facilities to compost it. SHAREWASTE.COM

Compost Now Based in North Carolina, USA, Compost Now collects compostable materials from community members and local businesses, turning waste into compost. COMPOSTNOW.ORG

Shared Earth A great community project that connects people who have land with people who want to garden or farm in the USA. SHAREDEARTH.COM

Lend and Tend Connecting those who can no longer garden with those who can. Worldwide. LENDANDTEND.COM

Local Tools Online directory to locate your nearest tool library. Worldwide. LOCALTOOLS.ORG

Big Bug Hunt Community project to warn gardeners when certain pests and diseases are in the vicinity. UK and EU. BIGBUGHUNT.COM

Plant names

Common names for plants can vary depending on region and country. In many cases the most widely used name for a plant is also its Latin name.

ANNUALS

Bishops weed/Bullwort – *Ammi majus*
Borage – Cool tankard, *Borago officinalis*
Calendula – English marigold/Pot marigold, *Calendula officinalis*
Carnation – *Dianthus*
Cosmos – *Cosmos bipinnatus*
Gypsophila – Annual baby's breath, *Gypsophila elegans*
Honesty – Money plant/Silver dollar, *Lunaria annua*
Larkspur – *Consolida*
Love-in-a-mist – *Nigella*
Nasturtium – Indian cress, *Tropaeolum majus*
Opium poppy – *Papaver somniferum*
Starflower scabious/drumstick scabious – *Scabiosa stellata*
Stock, 'Night Scented' – Evening stock, *Matthiola longipetala* ssp. *bicornis*
Sunflower – *Helianthus annuus*
Sweet pea – *Lathyrus odoratus*
Tobacco plant – Nicotiana

HERBACEOUS PERENNIALS

Agave – Century plant/American aloe, *Agave americana*
Ajuga – Bugle/Brown bugle, *Ajuga Reptans*
Anemone – Japanese anemone Windflower, *Anemone*
Aquilegia – Colombine/Granny's bonnets, *Aquilegia*
Armeria – Thrift/Sea pink, *Armeria*
Aster – *Symphyotrichum novi-belgii*
Astrantia – Hattie's pincushion Masterwort, *Astrantia*
Aubretia – *Aubretia*
Cornflowers – Blue-bottle/Bachelor's buttons, *Centaurea cyanus*
Echinacea – Coneflower, *Echinacea*
Echinops – Globe thistle, *Echinops*
Evening primrose – Coffee plant, *Oenothera biennis*
Foxglove – *Digitalis purpurea*
Geranium – Meadow cranesbill, *Geranium pratense*
Heather – *Calluna*
Hellebore – *Helleborus*
Hollyhock – *Alcea rosea*
Michaelmas daisy – *Aster novi-belgii*
Monarda – Bee balm/Bergamot, *Monarda didyma*
Peony – *Paeonia lactiflora*
Perovskia – Russian sage, *Perovskia*
Primula – Primrose, *Primula vulgaris*
Rudbeckia – Black-eyed Susan, *Rudbeckia hirta*
Salvia – *Salvia*
Scabious – Pincushion, *Scabiosa*
Sedum – Stonecrop, *Sedum*
Teasel – *Dipsacus fullonum*
Verbena bonariensis – Argentinian vervain/Purpletop vervain, *Verbena bonariensis*
Wallflowers – *Erysimum*
Winter aconite – *Eranthis hyemalis*
Yarrow – *Achillea*

BULBS

Allium – *Allium*
Bluebell – *Hyacinthoides non-scripta*
Crocus – *Crocus*
Daffodil – *Narcissus*
Dahlia – *Dahlia*
Grape hyacinth – *Muscari*
Hyacinth – *Hyacinthus*
Lily – *Lilium*
Ranunculus – *Ranunculus*
Snake's head fritillary – *Fritillaria meleagris*
Snowdrops – *Galanthus*
Tulip – *Tulipa*

CLIMBERS

Clematis – *Clematis*
Honeysuckle – *Lonicera periclymenum*
Hydrangea – Climbing hydrangea,
   Hydrangea petiolaris
Ivy – *Hedera*
Jasmine – *Jasminum officinale*
Passionflower – Granadilla, *Passiflora*

SHRUBS

Abelia – Glossy abelia, *Abelia*
Arbutus – Strawberry tree, *Arbutus*
Azalea – *Rhododendron*
Barberry – *Berberis darwinii*
Buddleia – Butterfly bush, *Buddleja*
Camellia – *Camellia japonica*
Castor oil plant – Japanese aralia,
   Fatsia japonica
Ceanothus – California lilac, *Ceanothus*
Daphne – *Daphne*
Dog rose – Bird briar/Briar rose/Wild
   rose, *Rosa canina*
Escallonia – *Escallonia*
Fuchsia – *Fuchsia*

Hebe – Shrubby veronica, *Hebe*
Lavender – *Lavandula angustifolia*
Mahonia – Oregon grape, *Mahonia*
Rhododendron – *Rhododendron*
Ribes – Flowering currant, *Ribes*
Sarcococca – Sweet box/Christmas box,
   Sarcococca
Skimmia japonica – Japanese skimmia,
   Skimmia japonica

HEDGING

Beech – Common beech, *Fagus sylvatica*
Blackthorn – Sloe, *Prunus spinosa*
Hawthorn – May tree, *Crataegus monogyna*
Pyracantha – Firethorn, *Pyracantha*

TREES

Acer – Maple/Japanese maple, *Acer*
Almond – *Prunus dulcis*
Amelanchier – Snowy mespilus,
   Amelanchier
Apple – *Malus domestica*
Cherry – *Prunus*
Hawthorn – May tree, *Crataegus monogyna*
Holly – *Ilex aquifolium*
Laurel – Cherry laurel, *Prunus laurocerasus*
Peach – *Prunus persica*
Pear – *Pyrus*
Pussy willow – Goat willow, *Salix caprea*
Tulip Tree – Tulip poplar, *Liriodendron
   tulipifera*
Willow – *Salix caprea*

HERBS

Basil – *Ocimum basillicum*
Chamomile 'Treneague' – Lawn
   chamomile, *Chamaemelum nobile*
   'Treneague'

Chives – *Allium schoenoprasum*
Coriander – Cilantro/Chinese parsley,
   Coriandrum sativum
Fennel – Common fennel,
   Foeniculum vulgare
Marjoram – *Origanum majorana*
Mint – *Mentha* ssp.
Rosemary – *Rosmarinus officinalis*
Sage – English sage, *Salvia officinalis*
Thyme – *Thymus vulgaris*
Thyme (Creeping) – *Thymus serpyllum*
   'Coccineus'

GREEN MANURE/
FERTILIZERS

Clover – Crimson clover/Italian clover,
   Trifolium incarnatum
Comfrey – *Symphytum officinale*
Mustard – *Sinapis alba*
Nettle – Common nettle/Stinging nettle,
   Urtica dioica
Rye – Hungarian grazing rye, *Secale cereale*
Vetch – Tares, *Vicia sativa*

HOUSEPLANTS

Corn plant – Dragon plant/Dragon tree,
   Dracaena fragrans
Devil's ivy – Golden pothos,
   Epipremnum aureum
Donkey's Tail – *Sedum burrito*
Earth Star plant – *Cryptanthus bivittatus*
Echeveria 'Blue Sky'
Jade plant – Money tree/Friendship tree,
   Crassula ovata
Nerve plant – *Fittonia*
Peace lily – White sails/Spathe flower,
   Spathiphyllum wallisii
Polka dot plant – *Hypoestes*
Rubber plant – Rubber tree/Rubber fig,
   Ficus elastica
Snake plant – Mother-in-law's tongue,
   Sansevieria trifasciata
Spider plant – *Chlorophytum comosum*
   'Variegatum'
Weeping fig – Benjamin tree/Java fig,
   Ficus benjamina
Zebra cactus – *Haworthia*